How To Be Your Own
POWER COMPANY

How To Be Your Own
POWER COMPANY

Jim Cullen

with J.O. Bugental

DRAWINGS BY GARY VERRALL

Technical Consultant, Clyde G. Davis

VNR **VAN NOSTRAND REINHOLD COMPANY**
New York   Cincinnati   Toronto   London   Melbourne

Copyright © 1980 by Jim Cullen
Library of Congress Catalog Card Number 80-13438
ISBN 0-442-24345-6 (paper)
ISBN 0-442-24340-5 (cloth)

Printed in the United States of America.

Published by Van Nostrand Reinhold Company
A division of Litton Educational Publishing, Inc.
135 West 50th Street, New York, NY 10020, U.S.A.

Van Nostrand Reinhold Limited
1410 Birchmount Road
Scarborough, Ontario M1P 2E7, Canada

Van Nostrand Reinhold Australia Pty. Ltd.
17 Queen Street
Mitcham, Victoria 3132, Australia

Van Nostrand Reinhold Company Limited
Molly Millars Lane
Wokingham, Berkshire, England

16 15 14 13 12 11 10 9 8 7 6 5 4 3 2 1

Library of Congress Cataloging in Publication Data
Cullen, Jim.
    How to be your own power company.
    Based on the author's The wilderness home power-
system and how to do it, published in 1978.
    Includes index.
    1. Storage batteries—Amateurs' manuals.
    2. Electric power production—Amateurs' manuals.
    3. Dwellings—Power supply—Amateurs' manuals.
    I. Bugental, J. O., joint author. II. Cullen, Jim.
Wilderness home powersystem and how to do it. III. Title.
TK9917.C84   621.31′2   80-13438
ISBN 0-442-24340-5
ISBN 0-442-24345-6 (pbk.)

The author has checked all of the systems described in this
book; however, he can take no responsibility for systems
installed by the reader.

The author has applied for registration for the following
trademarks: Wilderness Home Powersystem, Powersystem,
Wilderness Home Powercharger, Fastcharge Overide, and
Powercharger.

Designed by Jean Callan King/Visuality

For my friend Gary

My thanks to Orville Archer, Darrell Armstrong,
Mario Agnello, Mel Becker, John Bourn, John Bradley,
Craig and Nancy Brown, Larry, Sharon and Darrell Cechin,
Ethel Cullen, Susan Davis, Kathy Fennone, Aric Frasher,
Bernie Geiger, Edward Grazda, Verdan Goodyear,
The Hummels, Steve Johnston, Len Kuras, Jerry Martin,
Gladys and James Maxie, Mike from the mountain,
Christopher Neary, Gordon Nelson, John Newell,
Chuck Perelka, Helen Pettit, Milton Rice,
Gary and Phyllis Sargeant, Jim Scencenbaugh, Bobby and
Marilyn Thompson, Mary Verrall, Gary Wimmershoff, the
talented staff at Van Nostrand Reinhold Company and my
friends everywhere, from the mountains to the deserts, who
are proving it can be done.

Contents

Preface

How To Be Your Own Power Company presents a simple method of home electrification that is cost-efficient and easy to install. This book is based on The Wilderness Home Powersystem and How To Do It, which was published in 1978. The revision and expansion of that book, presented here, reflect some major changes taking place as individuals and whole communities throughout the world begin thinking about the supply and demand of energy, now and in the future.

Since the introduction of the first book, I have become personally involved in user-installed power for at least 200 homes in America's rural wilderness. Judging from the sales of the first book last year, there are at least 5,000 other homes considering the application of low-voltage power.

A typical home operating on my Wilderness Home Powersystem runs on 12-volt DC electricity about 90 percent of the time and on 110-120 (or 220-240) volt AC electricity the rest of the time. Both kinds of current are produced from a 12-volt-battery storage system.

One of the major changes in this new book is the application of new technology to the charging of those 12-volt batteries. The chapters "Charging Systems" and "Plugging In Your Home" have been greatly expanded to include consumer-oriented advances in wind plants, water turbines, solar arrays, and thermopile as sources of energy. My original system, dependent upon the automotive alternator in my car, is still valid but has also changed with the development of a fast-charge override device to control the battery-charging rate. The new device is now part of Chapters 4 and 8.

In addition to these changes in the body of the text, two more new developments are included among the appendices. One is an installation procedure for the Emergency Home Powersystem, an adaptation for city-dwellers who can no longer count on utility-company power as they once did. The other is a table of metric conversions, which can broaden the use of this book for a system where many measurements are involved.

All these changes respond to two growing trends in our society. One trend is the rapidly advancing technology that provides so many possible alternatives. The other trend is closely related: a rapidly growing interest among consumers, people like myself, who are looking now for alternatives to traditional utility-company power.

The National Electrical Reliability Council in

Princeton, New Jersey, has recently forecast an ever-increasing series of power shortages in the 1980s. The results will be not only brownouts but also total power failures over wide areas. New York City's fabled experience with blackouts is only a beginning.

Even as I write, something incredible is happening. The giant power company serving nearby towns has begun to buy costly TV air time for a new purpose. TV sponsors generally try to increase the demand for their products, but this power company is asking customers to consume less of their production. They are even threatening the use of brownouts and rolling blackouts to reinforce their requests for reduced demand.

I am not one of their customers, but if I were, I surely would resist their strange priorities and peak demand problems. I want to operate fans and air conditioners on a hot afternoon, not in the middle of a cool night. I want to operate kitchen appliances at meal times, not at the convenience of the utility company. Rather than accept the guilt they are laying on me, I have found a way to supply my own demand.

The problem with the traditional power company is its size. By planning for thousands or millions of people, it has lost track of realistic, individual needs. The same power-generating capacity that proves insufficient for peak demand periods still operates while we, the consumers, sleep.

Although hydroelectric dams have some storage capabilities, the big power-company system as a whole is wasteful. It has evolved in terms of maximum demands, maximum growth, and maximum consumption of limited natural resources.

The public-utility companies frequently burn precious fossil fuels to meet the grandiose demands of community power. One estimate says that utility companies deliver only one unit of electricity for every three equivalent units of dwindling earth resources that they consume. Even while they go on building bigger and more dangerous power plants to meet unrealistic demands, they ask for greater allotments of public monies and charge higher prices to individual users. And now they have the temerity to ask us to consume less power. This absurdity has to stop somewhere. For myself and a growing number of friends, it has already stopped.

Now that I am my own power company, I have custom-designed the supply for my own real demand. I also avoid waste in a system that stores almost all of the electricity it produces. With proper planning and use of low-voltage technology, my independent powersystem generates and consumes just the power needed for use now or later, and not one bit more.

An independently charged, battery-operated powersystem will serve you precisely because it is not too big. By combining new and old human technology with natural resources other than fossil fuels, you gain greater long-term security. And that is what *How To Be Your Own Power Company* and The Wilderness Home Powersystem are all about.

PART ONE
Design Your System

The Wilderness Home Powersystem

When I set out to make a lifetime dream come true, I had to give up valuable property in Los Angeles, complete with property taxes and monthly bills. I also gave up a fast-paced executive position, complete with headaches and stomachaches.

In their place I gained life here in the beautiful but harsh mountains of Mendocino County and the chance to build my own small home. I also gained some new neighbors, a special breed of people—many of whom have chosen to live better outside the same urban maze I left behind.

But there are many aspects of city life that I did not want to abandon, comforts and conveniences we take for granted. I honestly need my records, tapes, TV specials, and a reading light to keep my life complete.

Electrical power emerged as the greatest problem for me to solve in my wilderness home. I spent more than a year developing the first system to meet my needs.

My house sits on a mountain, 15 miles from the nearest town and four miles from public utilities. I know that it will be some time before they bring electricity to me, if ever. The cost to run a line just the four miles is close to $200,000 today, and the power company does not seem anxious to add new customers even at that price.

The solutions to my problem are the 12-volt technology employed in today's totally self-sufficient recreational vehicle (RV) and even some of the low-voltage applications developed for outer space. I began to power my mountain retreat by simply driving up and plugging the house into my car.

That was the first Wilderness Home Powersystem. Now there are almost infinite possible variations on the first system, plus adaptations for emergency home lighting in the city during power failures.

The second problem I had to face was how to support myself on a mountain. My powersystem solved this one as well. Many people were experimenting with 12-volt systems, but no one had explored in any comprehensive form the full scope of what could be done. I wrote my first book, *The Wilderness Home Powersystem and How To Do It*. I sold out the first printing by mail order in one year.

The book led to the opening of my little store in nearby Laytonville. It is a very unusual store, more like an experimental workshop with a lot of interesting new products in stock. My friends and I have become a clearinghouse for 12-volt technology and alternative-energy systems in general. We do not have all the answers, but we have begun to find out what works. In this new book, I want to show you how some of these ideas can provide a simpler, better way of life wherever your own lifetime dream takes you.

As I began working at my new jobs, I was also enlarging my home from a small two-room cabin to a house with more than 1,500 square feet—with two bedrooms, an art studio, a kitchen, a dining room, and a living room. As my home expanded, so did my electrical demands. I have been researching the 12-volt system to see just how far it will go. I flick a convenient switch for intermittent 110-120-volt AC power to run a food blender, mixer, and vacuum cleaner. I can even run a washing machine. When I turn the washer on, 110-120 volts of AC power are delivered automatically from my quiet, special inverter. My noisy, gasoline-powered AC generator runs only when I want it to, usually not at all, and I still have the extra power when I need it. The generator is merely a backup system.

When friends visit now, and I demonstrate my constantly changing system, they leave wide-eyed, thinking of all the possible combinations in their own homes. I never operate the AC generator when I have company, because I do not want my friends to hear its loud engine. Neither do I depend on power lines strung around the beautiful rolling hills or across the creek near my house.

I am truly self-sufficient on my mountain. Nothing short of a major earthquake can knock out my power. In fact, not too long ago, California weathered one of the most violent windstorms in recorded history. To the south, Bakersfield was totally blacked out. Eureka and Humboldt Counties, north of here, and most of Mendocino County, where I live, had almost total power failure.

I even felt good—but more than a little concerned—as the TV news announcer from Eureka said, "If you are one of the few people able to watch us, as you obviously are, you are one of the few to have power. . . ."

The next time the lights go out, I want you to join me as "one of the few." But why wait for the storm? You can begin to relax with security and city conveniences in your wilderness home right now.

Whether you drive a car, a pick-up, a Jeep, or whatever, your first system can be right there in it. We will refer to it as the car, for the sake of simplicity, from now on.

In my original two-room home, every time I started my car and drove to the store or to a friend's home, the alternator recharged the battery. The cost of powering my home was part of the fuel bill to operate my car.

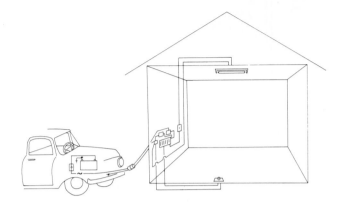

I could get two and sometimes three days' use from my system before I had to start the motor to charge the battery. If I had nowhere to drive, I ran the engine slightly above idle speed until the battery was charged.

When I first built my bigger home, I installed a dual-battery system. The same industrial-quality batteries that drive forklifts and electric cars now power my home. For others, I have designed and installed systems that utilize gasoline-powered generators already on the property—but with reduced operating costs and great increases in efficiency.

If you do shop work as a hobby, want to use a washing machine, or have any need for high-wattage 110-120-volt AC power for extended periods, you can charge your home system while using your traditional generator, gasoline powered or converted to liquefied petroleum gas (LPG) or propane, as seldom as once a week. Your batteries will be fully charged and supply one week's 12-volt power for your home while you work at your hobby or whatever. This can save you from 50 to 75 percent of fuel and maintenance costs.

Either your automotive alternator or your portable generator can charge your 12-volt powersystem. So can the energy of the wind or a water supply in sufficient quantity. So can the light from the sun or heat from a thermopile source—or a combination of any two or more power sources.

In Chapter 4 "Charging Systems," we will cover these alternatives in more detail. But please keep them in mind as you read about other aspects of the system. You can build it to meet your own needs and your own existing equipment.

What exactly does the 12 volt system offer? In my home, it provides room lighting equal to that in a city home. I also operate TV (black-and-white or color), big sound stereo, and electric razors. In addition, I have a pressurized water system that is totally automatic. Without looking closely, you would never know the house is not operating on utility-company power.

The number of 12-volt accessories on the market today will amaze you. Everything from an automatic coffee maker to an iron—even a home computer—is available. With the addition of a solid-state inverter or a rotary inverter, the possibilities reach unlimited proportions. A solid-state inverter normally must be switched on or off by hand. A rotary inverter can be controlled by the appliance it serves. These are just a few of the choices you will make from the list of appliances and accessories that follows in the next chapter.

I want to emphasize here that the Wilderness Home Powersystem is not a panacea. There are certain appliances and accessories the system cannot handle efficiently—unless you are blessed with sufficient water, wind, or sunshine and have the monetary means to tap them. I mean things like big refrigerator-freezers, electric heaters, induction motors, air-conditioning units, etc.

You can do without many of these appliances if you build or convert your home with the elements and situation in mind. Although there are 12-volt refrigerators, I use propane gas to run mine. I use wood for heat, and my house has big glass windows facing south to catch the sun and lots of insulation to help maintain a comfortable temperature throughout the year.

As part of this total plan, my 12-volt powersystem provides solutions to most of my needs. Reaching a plateau now in the development of the system, I realize that it really has not been hard to put together. If a former executive with minimum technical experience can do it, any practical do-it-yourselfer surely can.

I can follow simple logic about as well as most people, but when something is written by a technician, I can usually forget it because of its complexity. That is why I have written *How To Be Your Own Power Company* with people like myself in mind. I have avoided, as much as possible, digging too deeply into electrical theory. My goal is to offer a simple design and installation procedure in familiar terms.

But there are some electrical terms and other words you will see over and over again as you become involved in building your own system. They are common words and abbreviations stamped right on the components you will be shopping for. To avoid these terms would only complicate matters for both of us, so I use them and explain them as I go. So you can always look them up easily, Appendix A is a glossary of these terms.

I know, for many people, the minute electricity and wiring are mentioned, all kinds of complicated visions come to mind. There are fears associated with shocks and danger we have learned to expect from regular household current.

One beauty of the 12-volt system is safety. There is little or no shock hazard, usually just a slight tickle. However, electricity in any form, when installed improperly or through shortcuts, can pose a fire hazard. That is why I have included in Appendix B an interpretation of applicable sections from the National Electric Code. Most places where we build are governed by an electrical code of one sort or another. Check the local building permits office in your city, county, borough, or parish. If no code requirements exist, follow the National Code. It is not difficult and does not add much to the system's cost. The bottom line is peace of mind.

I have found some people react negatively when I

tell them that a 12-volt battery is at the heart of the system. If you are anything like I used to be, you do not become aware of a battery until it goes dead. When it does, it becomes a glaring problem for a short moment.

No wonder some of my friends were skeptical when I first told them about my 12-volt system. One thing I heard was, "If I leave my car lights on all night, my battery will be dead in the morning, so how can I run a home on it?"

The answer is that most people do not know how powerful the car headlights are—and how much current they draw. When the headlights are on, the tail lights, parking lights, side lights, and dash lights are also all illuminated. Without going into any technicalities, I can tell you that there was enough power in a car battery to run my first two-room cabin for two days. And I still had enough reserve power to start my car.

In earlier years, the battery was considered good for cranking up the car, but that was about all. It could run down or discharge rather quickly, because it was the heart of a six-volt system. Accessories required tubes and were inefficient.

Manufacturers were soon forced to introduce the 12-volt system, capable of meeting much higher power demands. With air conditioning, car TVs, high-intensity lighting, coffee pots, and who knows what-all plugged into the cigarette lighter, what other choice was there?

Then came the travel trailer, which is virtually a complete home and is plugged into the family automobile, too. Then came the super motorhome, then the RV revolution. By this time, of course, the old generator in the car had become as obsolete as the biplane. The alternator became the standard battery-charging unit, because it could handle the new power requirements so much better and recharge the batteries much more quickly. Space-age technology blasted its way into the RV picture, too, as it has into practically every phase of our lives.

Travel trailers and motorhomes today are complete homes-away-from-home with every convenience imaginable. Transistors and miniaturized circuitry have made a lot of conveniences more portable. Many RVs even have voltage reducers that convert the 110-120-volt AC at a trailer park down to 12 volts.

Concurrent with the RV demand and supply, heavy industry contributed to battery development, too. Electric forklifts and golf carts are just two examples.

If the 12-volt battery works so well for motorhomes, travel trailers, and industry, it will work for our wilderness homes, or even in the city, just as well—with proper planning.

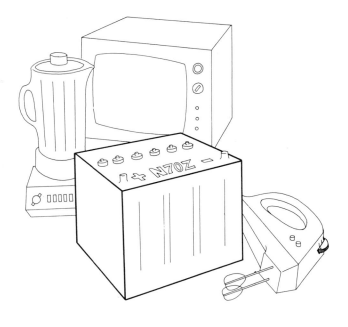

Now that you know a little more about our friend the battery, here is a suggestion I think will be fun and help you understand its function.

Visualize the battery you will choose as a checking account. You can make withdrawals from your account as the need arises. Only instead of money, you withdraw electric power. The missing link is a long-lost relative with independent means, someone who can make regular deposits to your account. Your automotive alternator is one such wealthy relative. It can deposit up to a week's worth of power in your account in just a few hours. But in these inflated times, we are fortunate to have any number of wealthy relatives—alternative charging systems—to keep our accounts filled.

The amount of power you may withdraw from your battery will be governed by its capacity, by how much power the appliances you want to operate will use each day, and by how long and at what output you operate the charging system in order to replenish your account.

The result of all this is a Wilderness Home Power-system of your own. First, we will discuss how to plan and install the simple system. By starting simply, you will gain an understanding that will keep pace with the growth possible in your system. Then we will discuss some flexible options and how to power a much bigger home with many household conveniences. Finally, we will talk about the limits of the system and where it can go from here.

I have been having a lot of fun with my experiments and variations on this powersystem. As more and more people continue to install, use, and develop it to meet their own particular needs, we are all learning from each other. You may run across applications of the system I have missed. Conversely, you may run across problems I did not encounter. In either case, I would like to hear from you. There is no telling where we can go with our combined experience.

We have talked long enough about my system. Now let us get to work on yours. The best system for your home will result from two equally important basic procedures, your design and your installation. Part Two of this book presents step-by-step instructions to install your system, but you need to give careful attention first to what sort of system yours will be. In Part One, you must make a number of decisions—what your needs are, which is the best battery, and which of the many charging systems will work best on your property to meet your own custom requirements.

If you are ready, let us begin at the end, to develop an idea of your finished product.

2
Making Your Choice

When you move to a home in the wilderness, you can take many of the conveniences of city life with you. Which of them do you want?

Knowing the end result, your own consumer demand, is necessary before you can choose the size of batteries for your home and develop the appropriate charging system to keep up the power supply. You must know what appliances and accessories you will want to use in your home and how much power those accessories will consume per day. Then you can compare your probable use to how much electricity your system can store. These considerations form the first step as you design your own powersystem. They are all your choices.

Just as you have planned the shape of your house in the mountains, on the desert, or by the sea, take time now to think about how you will spend your time. How will your family spend theirs? What are all the ingredients of an especially good time for you and your friends?

For me, it is a treat to come in out of the woods and turn on my big-screen color TV when a special program is aired. My friends and I also get a kick out of Sunday brunches served as we whip up frozen daiquiris in my kitchen blender. One night we followed a special dinner of venison in wine sauce with Grand Marnier soufflé. We rely on the electric mixer. Homemade cake, anyone?

For some of my neighbors, the question is how much time they actually will be operating the power saw or the drill press. Some need to think about the sewing machine, soldering iron, or wood-routing tools. We all have to think about the vacuum cleaner once in a while.

On the next few pages, I have listed all of the readily available 12-volt appliances I found. These particular components are reliable and easily adaptable to home use. But 12-volt technology constantly becomes more refined. Many new products may be added to the list when you begin to shop.

Take a pencil now, and as you think through typical days at home, mark the items you will want to include. Following the description of each product is an approximate price and a code for the supply source. Some of the most likely sources for each component are noted by initial codes as follows.

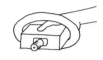

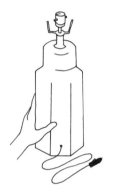

RV   Recreational vehicle or camper supply.

ELEC Electronic supply house, communications specialists. See Yellow Pages of your phone book. (Radio Shack is the widest based operation and also has an excellent catalog of equipment compatible with the 12-volt system. If there is no dealer near you, you can ask that your name be added to their mailing list by writing to them at 2617 West 7th Street, Fort Worth, TX 76107.)

MRNE Marine supply store.

MAIL Mail-order catalogs. The last page of this book is a form for The Wilderness Home Power-system Supply Catalog, Box 732, Laytonville, CA 95454. Another good company with a catalog is the automotive supplier, Warshawsky & Co., Box 8440A, Chicago, IL 60680.

AUTO Auto supply store or car dealer, service and parts department.

HDWE Hardware store.

You can also receive more information about many of the items listed here, and even newer equipment not yet listed, by writing to me in Laytonville. I hope that you will want to order the catalog described on page 143.

The final columns in the following lists indicate the approximate electrical current drain for each item in amps. More or less than one whole amp has been denoted by a rounded-off decimal figure. When you multiply the actual length of time you will use an appliance (in hours) times the current required to operate the appliance (in amps), you will know your particular electrical demand in *amp-hours*. After you have compiled your list, we can add all the figures to compute the total amp-hour demands in your home.

12-Volt Conversions

Many lamps and small accessories can be converted directly from 110–120-volt AC electricity to the 12-volt system. Just replace the existing plug with a 12-volt cigarette-lighter-type plug or an AC plug adapter. Then there are two types of receptacles for the new plug.

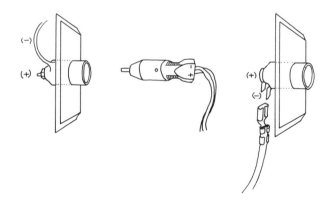

Each new plug will cost under $1.50. Plug adapters are about $1.75. The matching receptacle on the left costs about $3.25. All are available from RV and MRNE. The receptacle on the right is a new design, explained in my catalog. "Its" installation is covered in Chapter 10.

Ordinary Lighting

There are 12-volt bulbs available that look exactly like the bulb in a normal household lamp, but you can only use them with your converted lamps. If you try to use them in lamps with 110–120 volts, they will burn out. To be sure what type of bulb you have, always look at the top of it for specifications on the voltage and wattage.

Fluorescent Lighting

For maximum lighting output, consider 12-volt fluorescent lights as an alternative to conventional lighting wherever you can. They draw as little as one fifth the power of conventional lighting and shine four to five times brighter. Also ask your dealer for fixture brands, such as McLeans, which use regular fluorescent tubes

Flourescent Lighting

you can buy almost anywhere. You can expect a tube to last from 2,000 to 4,000 hours in a 12-volt system.

You cannot buy a regular 110–120-volt fluorescent fixture and plug it into a 12-volt system. These fixtures actually have miniature inverters in them that step 12 volts up to 110–120-volt AC to excite the lighting tube.

15-watt, slimline fixture. About $25.	RV, MRNE, MAIL	1.0
16-watt, slimline fixture, dual 8-watt tubes. About $20.	RV, MRNE, MAIL	1.2
30-watt, slimline fixture About $30.	RV, MRNE, MAIL	2.0
22-watt, slimline fixture. $25-30.	RV, MRNE, MAIL	1.4
Mood light, black/white fluorescent for posters, etc. $30-35.	RV, MAIL	1.4
22-watt, circleline, lamp conversion, single tube. $35.	MAIL	1.5
22-watt, circleline fixture, single tube. $30.	RV, MRNE, MAIL	1.5

Communications

An adapter, available at most electronic stores or by mail order, will allow you to plug into the 12-volt system your existing 6-, 7 1/2-, or 9-volt accessories—including TV games and many home computers. You can also plug 6-volt calculators into the system to charge the nickel cadmium batteries on which they operate. Adapters are priced from $10 to $15.

Black and white TV, 9" or 12" screen. $90-150.	RV, ELEC, MRNE, MAIL, HDWE	1.1
Color TV, 9" screen. $330-400.	RV, ELEC, MRNE, MAIL, HDWE	4.0

Digital clock with LED readout and alarm. $40-60.	RV, ELEC, MRNE, MAIL, AUTO	0.1
CB radio. $65-500.	RV, ELEC, MRNE, AUTO	Receive 0.3 Transmit 0.5
CB scanners. $100-500.	ELEC, MAIL, AUTO	0.3
Radio/telephone (FM). $500-2800.	ELEC, MRNE	Receive 0.3 Transmit 2.5-15.0
Auto stereo: FM/AM, 8 track and casette. $40-300.	RV, ELEC, MRNE, MAIL	0.5
Booster amplifier-equalizer 30 watt (15 per channel) 40 watt (20 per channel) $40-125.	ELEC, MAIL, AUTO	2.0 3.0
60 watt (30 per channel) 100 watt (50 per channel) $125-300.	ELEC, MAIL	4.0 5.0
DC turntable (converted to 12 volts)	ELEC, MAIL	.5

Tools

Air compressors: inflate toys, air mattresses, footballs, etc. $20-45.	RV, MRNE, MAIL, AUTO	3.5-9.0
Winches and related heavy-duty motors. $110-600.	MAIL, AUTO	10.0-100.0
Chain saw, 12 volt, 10" or 14" blade. $150-160.	MAIL, HDWE	100.0
3/8" drill. $45-55.	MAIL, HDWE	12.0-14.0

Safety

Propane gas and smoke alarms. $25-35.	RV, MRNE, MAIL, HDWE	0.3
Intrusion alarms—no power draw unless activated, usually 85 decible bell or siren. $25-35.	RV, ELEC, MRNE MAIL, HDWE	0.3

Hygiene

Electric razor converter, any razor to 12 volts. About $20.	RV, MRNE, MAIL, AUTO	1.3
Vacuum cleaner (hand) $10-40.	RV, MRNE, AUTO	3.5-22.0
Travel iron. $10-15.	RV, MRNE, MAIL	10.0

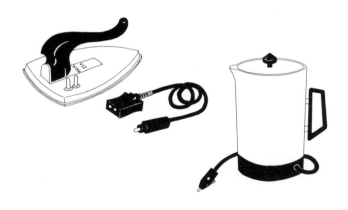

Kitchen

Coffee percolator—two-six cup, 12 volt and 115 AC. $10-30.	RV, MRNE, MAIL	11.3
Vent fan—suitable for bath or kitchen exhaust, heat circulation— 5" blade, 2600 RPM. About $15.	RV, MAIL	0.5-2.5
Range hood—with light & fan. $30-45.	RV, MRNE, MAIL	5.0
Toaster.	RV, MRNE, MAIL	20.0
Popcorn popper.	RV, MRNE, MAIL	16.0
Slow cooker. $25-35 each.	RV, MRNE, MAIL	20.0
Mixers. Blenders. Can openers. Food processors. Knife sharpeners. Electric knives. Coffee grinders. Meat slicers. Microwave oven— 1,000 watts max. Dishwasher— heating element bypassed.	See **Inverters** list for intermittent AC power required for these appliances.	

Inverters, Solid-State and Rotary

A 12-volt system alone will take care of many of my basic needs. But now that I live full-time in my expanded wilderness home, I want to run such things as my 19-inch color TV, a sewing machine, and a multitude of kitchen accessories that require AC current. Many AC appliances can be used on the 12-volt DC system—with the help of a device called an inverter. We will discuss inverters here because they are accessories on the 12-volt powersystem. They take energy to run themselves, ten percent or more of their capacity, plus the energy the appliance draws. You need to shop around for the most efficient inverter if you choose one as a component for your own system.

I do not want to get involved in the mathematics and physics of inverter operation, but a few points will be helpful. Basically, the inverter takes nominal direct current (DC) from your battery-storage system and changes the waveform to alternating current (AC) at a particular voltage and frequency. There are technical names for the different forms, square wave or sine wave, and there are synchronous and non-synchronous types of inverters. Let us concentrate on applications rather than theory, though.

Synchronous-inverter systems provide DC-to-AC conversion in heavy-duty industrial situations and certain unique home applications, all of which depend on the power capacity of a public utility company. They cannot use battery storage, and at this point, they are very costly. Technology may change some or all of these facts in the next few years, but until that time, synchronous inverters have no place in a discussion with those who want to be their own power companies.

Square-wave, non-synchronous inverters can have their place in our system, but most will not run the induction motors of air conditioners or refrigerators. An iron, toaster, or heater that exceeds the inverter's wattage capacity cannot be used either. These appliances produce heat by allowing a very large current to flow into a low-resistance heating element, causing power (heat) losses in the wiring to the appliance—most wasteful in terms of energy used. Inverters themselves are not intrinsically very efficient, and if left on continuously, they will drain the batteries more quickly than even high-amperage 12-volt accessories. But if we use them correctly, that is intermittently or only when needed, we can justify the inefficiency by the flexibility gained in a 12-volt powersystem.

As an example of the luxury inverters offer, one Christmas we decorated all the windows of the house with miniature colored lights—in the middle of the wilderness. We left them illuminated for about three hours each night, using up only about nine amp-hours.

A number of very low wattage inverters are also available. They are practical in specific applications, for instance with a low-wattage sewing machine or even a console color TV. Just make sure the wattage on the specification plate of the appliance does not exceed the wattage capacity of the inverter. If it does, the inverter will blow a fuse or even burn out.

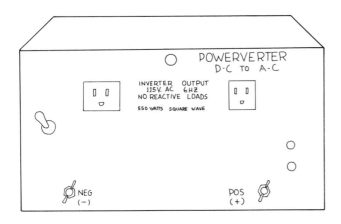

For our basic purposes, an intermediate inverter of 500 to 1,000 watts is ideal. It seems to be the most efficient for operating the wide range of light household equipment requiring AC power, things like those listed here under the "Kitchen" heading, including the dishwasher (without a heating element) or a microwave oven that requires no more than 1,000 watts. Among inverters up to the 500-watt size, look for silicone transistors to assure long durability. But in higher wattage units or some special applications, germanium transistors may serve better because they insure lower power loss.

Here are some of the solid-state inverters available. All but the last two produce square-wave current. The current in amps listed here is about what these models draw on idle, not under a load.

100 watt, no polarity protection. Under $50.	RV, MRNE, MAIL, AUTO	1.0-3.0
250 watt, no polarity protection or frequency control. Under $100.	RV, MRNE, MAIL, AUTO	2.5-3.5
400 watt, no polarity protection or frequency control. About $110.	RV, MRNE, MAIL, AUTO	4.0-5.0
550 watt—500 continuous, 600 intermittent. Ask for polarity protection and frequency control for operating TV, tape and similar equipment. Also operates 3/8" drill and other small tools. $140-390.	RV, MRNE MAIL, AUTO	6.0-7.0
1,000 watt—continuous. Same as above but usually has frequency response controls. $425-750.	RV, MRNE, MAIL, AUTO	10.0-15.0
1,000 watt—continuous. Special modified sine wave. Model has 5,000-watt surge capacity for starting heavy-duty motors, refrigerators, freezers, etc. Frequency control. 90% efficient. $1,000-1,150.	MAIL	1.5

The need for solid-state, sine-wave inverters with high wattage capacity is generally limited to very specialized homes or workshops. They require constant charging, which probably will be dependent on the presence of abundant wind, water, or sunlight as energy sources.

For the most part, I do not recommend many of the inverters whose output exceeds 1,000 watts. They are expensive and draw heavily from your battery just to run themselves. An inverter rated for 3,000 watts will likely draw an idling current of 300 watts, enough to run my whole house for nearly four days. If I simply forget to flip off the switch for equipment of those dimensions, my battery-storage system would be wiped out. The cost is very high, too, up to $2 per watt.

The largest inverter I recommend is the modified sine-wave unit listed. Its output is 1,000 watts with 5,000 watts of surge power to start heavy-duty motors. You can get more information from my catalog or from Aero Power, 2398 Fourth Street, Berkeley, CA 94710.

When you have decided on an inverter, there are some optional features to look for.

Mechanical timers (0 to 60 minutes) can be quickly installed to insure automatic shutoff. They are available from many electrical supply dealers for about $10 to $12.

A demand-load switch is one protective device for any inverter installation. This automatic sensor only turns on the inverter when the load demanded by another appliance appears on the AC circuit, then it switches the inverter off when the load is withdrawn. One such device is included in my catalog.

Over-current protection will shut down both the inverter and the appliance drawing the current load when they exceed the maximum wattage levels. Avoiding a burnt-out inverter more than offsets the temporary inconvenience—especially important with this equipment in which your initial investment can really add up.

Voltage-frequency regulation is especially important with your sound equipment. Even minor fluctuations in operating speed can terrorize a good turntable and cause considerable wow and flutter on your favorite stereo albums, not to mention tapes. Unfortunately, frequency control can practically double the price of even the most reasonable inverter.

My friend Clyde, an electronics engineer, solved this problem for me. I sold my AC stereo equipment and bought a 12-volt system for tapes, radio, and even TV. Then I found a good magnetic turntable equipped with a DC motor, and Clyde converted the motor from 110-volts DC to 12, then built a 12-volt pre-amp to reproduce the sound from the magnetic cartridge. Now I not only have a total sound system, producing 50 watts per channel and customized for my wilderness home, but I also can save even more energy in terms of current drain by the equipment.

Sanyo, Panasonic, and Sony are among the manufacturers now offering complete 12-volt systems with phonograph jacks. Check to make sure the system you choose will reproduce from a magnetic cartridge. If not, you will need a 12-volt pre-amp. If you cannot find one, write to Clyde at C. Davis Avionics, P. O. Box 725, Laytonville, CA 95454.

The installation of an inverter in your home is a 110–120-volt AC installation. That means wiring for the inverter's output must conform with the local code in your area, probably under the supervision of a qualified electrician. You absolutely cannot use the same wires to carry 12-volt and 110–120-volt power at the same time.

For a small inverter of 200 watts or less, an alternative installation is to place the unit in the kitchen or a room where you figure it will get the most use. Then plug the appliance directly into the inverter. In this case, there is a noticeable hum or buzz during operation. That distracted me, so I decided to locate the inverter outside the house with a remote switch inside, right next to the blender, then did the necessary code wiring.

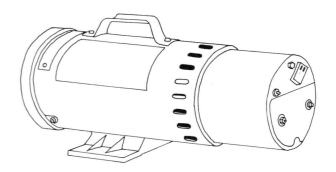

An alternative to solid-state inverters, the rotary inverter (also known as a motor-generator) performs essentially the same function of changing power. Its motor accepts 12-, 24-, or 36-volt DC input to drive a small, integrated generator, which produces 110–120-volt AC output. You may want to explore the 24-volt model if you need to use high-wattage power tools or other AC equipment over extended periods of time. But for our basic 12-volt powersystem, there are two sizes of rotary inverters, one with 500- and one with 1,600-watt capacity.

There are both advantages and disadvantages when you use the rotary rather than the solid-state inverter. Among the advantages: Both 12-volt models available have automatic demand switches to turn themselves on or off with appliance demand. Rather than a square-wave form, rotary inverters produce the sine wave required for sensitive electronics equipment. A rotary inverter will also start some motors that demand high initial power, but only to a 50 percent increase over these motors' ratings.

The biggest disadvantage of the rotary inverter is that it is a mechanical rather than a solid-state device, which usually means more maintenance and a shorter life span. The 60-hz frequency stabilization required by some sound systems is missing in rotary inverters. And their initial purchase price is high. Looking at a solid-state unit of comparable capacity, the current drain just to run a rotary inverter also seems disproportionately high.

A rotary inverter is relatively quiet but not silent. If the sound of its operation bothers you, you can buy compact models that can be located in a protective box right outside your house.

I know of only one company in the United States that now makes the 12-volt rotary inverter—Redi-Line, a subsidiary of Honeywell. Here are the two models currently available. Remember that the amp drain listed is based on the unit operating with no load.

500 watt—continuous, 750-watt intermediate, sine wave, automatic load sensing, model D12A. $375-400.	RV, MRNE, MAIL	0.0
1600 watt—continuous, 2,400-watt intermediate, sine wave, automatic load sensing, model DA12L. $750-800.	RV, MRNE, MAIL	0.0

For the present, the disadvantages of rotary inverters appear to outweigh the advantages. A good alternative is to install two or more solid-state inverters. I now use the unit with 1,000 watts and 5,000 watts surge capacity to operate the kitchen blender, washing machine, and mercury vapor lighting outdoors. The lighting is connected to a mechanical timer. The same inverter will permit me to operate a microwave oven or an electric dishwasher with its heating element disconnected. But I still use my smaller solid-state models for appliances that demand fewer than 500 watts—the hand-held mixer, the can opener, and the portable sewing machine. There is no need to start up the 1,000-watt unit for them.

Again, let me remind you about 110-120-volt AC circuits in your home. That is the current produced by the rotary inverter, too. Like the solid-state inverter, the wiring here must meet your local code requirements.

Water Pumps

All of us know the importance of water, but we tend to take it for granted. Of course, if you plan to raise animals or irrigate crops, as some of my neighbors do, you will think of water much sooner. But for me, the necessity was not really driven home until I found myself trying to cope without a city-type water system.

I built my home for the best view, and being new to country life, I thought it would be easy to solve my water problem. After all, there was a stream, and I had sunk a well. I also had an AC generator on the property. So my pump turned the generator on and off several times every day to deliver water whenever needed.

I guess that was a good learning experience, one reason I am now such a believer in my controlling the use of the generator. The generator simply would not hold up in that kind of constant on-off application, coupled with the rest of my AC demands. Carrying water back and forth in buckets while repairing the generator was no fun, either.

Then I took another look at the RV accessories. I found a wide variety of 12-volt water pumps, everything from a small one that turns on with a switch next to the faucet, to my present pump which provides pressurized water on a multi-fixture basis at $3\frac{1}{2}$ gallons per minute. It is totally automatic and operates only when a faucet is turned on.

I also use a manually operated 12-volt pump, actually a marine bilge pump. Drawing only about $1\frac{1}{2}$ amps, it serves a sprinkler system I fashioned from 50 feet of perforated garden hose. I can leave it on for several hours without a problem and preset my mechanical timer to shut it off.

To solve my water problem completely, I built a 3,000-gallon redwood holding tank. In winter time, the tank fills continuously with water from a stream above the house. During the summer, the stream dries up, and I use a five-horsepower engine with a fire pump to bring enough water from my well up to the tank. The 3,000-gallon supply can last me up to a month at a time.

Of course, I could have elevated my water tank above the house to provide indoor water pressure by gravity, but there were a few problems. The first was the expense. The second was that I would no longer be able to feed the tank by gravity from the stream, because the

Storage Water System

tank would be too high. Another alternative would have been to build the house closer to the well, but I did not want to live in a gully—not with the magnificent view I have now. So I chose the 12-volt pump for $150.

Some pumps have an external AC motor mounted on them. You may be able to replace that motor with a DC unit to run directly from your 12-volt battery system, saving both money and power. In many cases, you can also convert washing machine motors. The DC pump motors operate 1/8 to 7 horsepower and 200 to 5,000 RPM. When you consider one, be sure you explain to the supplier your requirements, including horsepower, RPM, and a description of the AC motor you want to replace. One good source is Applied Motors, Inc. You can call or write to them at P. O. Box 106, Rockford, IL 61105. Phone (815) 397-2006.

Many different 12-volt pumps are available from RV and camper supply stores. After experimenting with a number of pumps that failed under constant use, I found two that work pretty well, the Aquarius by Coleman and the Jabsco by ITT. The Jabsco has a belt drive, which will need replacement from time to time. The Aquarius does not.

About 3.5 gallons per minute, multi-fixture and automatic demand. $80-150.	RV, MRNE	6.0-9.0
3.1 gallons per minute, non-automatic—requires wall or faucet switch—good for multi-fixture flow. About $35.	RV, MRNE, MAIL	3.0
2 gallons per minute, non-automatic and single fixture only. Under $20.	RV, MRNE	1.5

All these pumps mount at surface level and can be housed right along with your on-site auxiliary battery, your inverter, or any other equipment. Details on the installation of a 12-volt pumping system are spelled out in Chapter 12.

There are also deep-submersible pumps designed to be mounted inside a well casing, but these pumps are generally 110-120-volt AC. The conversion of their motors is difficult because of their size and shape, and

because they are integrated or flange-mounted to a narrow frame suitable for sinking underwater inside the well.

No matter how good your well is, if you do not have gravity feed, I would encourage you to erect the largest possible holding tank. That way, you can limit the use of an AC power source for pumping and, at the same time, plug in the battery charger we will discuss in Chapter 4.

I know several people who use wind energy to pump water from their wells directly. Given sufficient wind and a big enough well, a pump 12 feet in diameter can move up to 415 gallons an hour in a 15-mile-per-hour wind. For further information, write to some of these water-pumping windmill manufacturers.

- Aero Motor Water Systems, Broken Arrow, OK 74012.
- Dempster Industries, Inc., P. O. Box 848, Beatrice, NE 68310.
- O'Brock Windmill Sales, Rt 1-12th St., North Benton, OH 44449.

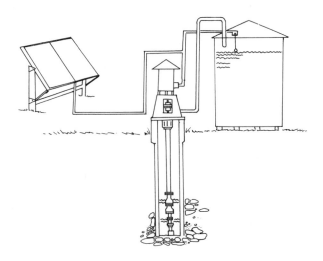

One of the world's most rapidly developing methods for pumping water involves the use of photovoltaics—electricity from sunlight. Solar electric pumps are expensive and most prevalent now in experimental community water systems, mainly in underdeveloped countries. Many do not use battery storage but convert 90 percent of their electrical output directly into pumping power. Research shows that a mere 600 peak watts of solar electricity can deliver up to 38,290 gallons of water a day. At this point, all photovoltaic pumping systems are custom designed, and price estimates begin around $4,000. For more information, write to Solarex Corp., 1335 Piccard Drive, Rockville, MD 20850.

Refrigeration

Most 12-volt refrigerators are not yet very practical on our systems because their absorption-type systems must operate constantly. The smallest units available have two to four cubic feet of cold storage space and draw a minimum of two amps or 48 amp-hours a day from the battery. I use an economical propane-gas refrigerator in my home, and its capacity is ample.

Employing the Peltier effect, which converts heat to cool air through solid-state modules and a small fan, Koolatron Industries may produce one of the most efficient 12-volt refrigerators. Their three-cubic-foot model will keep 40 pounds of food cold while drawing two amps. Two miniaturized thermoelectric modules replace all the complex coils and motors, except for a small fan, which is the only moving part in this unit. It is listed in my catalog, or you can write directly to Koolatron Industries, 56 Harvester Avenue, Batavia, NY 14020.

There are several refrigerators on the market today that have three-way switches for 115-volt, LPG/propane, or 12-volt operation. The selection allows you to switch to whatever kind of power is available, especially in an emergency. The 12-volt phase can work well for several hours whenever you run out of gas or if there is a problem with the gas system.

12-volt Koolatron, uses Peltier effect. Approximately three cubic feet. Chest type and built-in. Freezer available. $160-290.	RV, MRNE, MAIL	2.0-4.0
Three-Way Refrigerators, 12 Volt/AC/LP Gas: 2 cubic feet, 4 cubic feet, 7 cubic feet, 10 cubic feet. $300-995.	RV, MRNE, MAIL	2.0-20.0

Only a system that provides 12-volt energy around the clock will solve our problem with electric refrigeration. Solar energy alone could do it today, but only at great expense. If you live in an area with the right climate, wind or water generation could make electrical refrigeration practical now. But for most of us, gas is currently the most economical standby.

Heating

Your 12-volt system cannot provide direct heat. Electrical heating in general consumes enormous energy and is sadly wasteful. But the 12-volt system can operate fans to circulate heat produced by gas or wood.

RV technology has given us several makes of furnaces that burn LPG or propane. They come equipped with low amperage, thermostatically controlled blowers. They are smaller than city home furnaces and use less gas, but they produce 10 to 40,000 BTUs of heat. You can choose a furnace that suits your needs at the RV or camper supply store.

While you are looking at furnaces, be sure to consider the whole installation. Just two room registers are normally supplied with these furnaces, and you may want to order more. Also look for a model that attaches to standard four-inch duct. The standard duct is available at most sheet-metal works or heating-supply dealers, and it is easy to install.

New styles of gas-operated catalytic heaters, now thermostatically controlled, have recently been introduced. Safe and efficient, they are available from RV and camper supply stores or by mail order. Units that produce 12,000 BTUs of heat cost around $250.

Another heating alternative is the old standby, a wood stove. Be sure you select one of the newer models that have been engineered for maximum efficiency. I have used a fantastic wood stove fashioned from an old submarine net buoy by a local stovemaker, and it is large enough to heat several rooms at once—besides being a conversation piece. As if that were not enough, my friend Clyde recently developed a way to circulate the heat from my big stove. He added a small 12-volt fan to move up to 450 cubic feet of hot air per minute. The fan consumes only about 2 amps as it operates, and it turns itself on thermostatically. A thermal switch acts as a ther-

mostat and activates the motor when the stove reaches 140°F. The motor-driven fan then circulates the warm air.

I use equipment manufactured by Dayton and available through Grainger dealers. The motor, model #2M272, costs under $15. The fan is model #4C472 and costs $1.25. The thermal switch, automatic fan control model #2E250, is around $10. If you cannot locate a Grainger dealer, write directly to Dayton Electric Manufacturing Company, Dept. TR, 5959 Howard Street, Chicago, IL 60648. I have also put together a kit that contains the motor, switch, and asbestos or Teflon wiring, available through mail order.

Simple passive solar design provides an obvious way to capture heat for the rooms of your home. If you are building a new place, think carefully about exposure, shades, and insulation. If your house is already built, a little creative remodeling can go a long way toward temperature control—and you might even qualify for tax writeoffs, like those available in California.

When you apply passive solar principles, you begin with the sun's relationship to our planet's equator—right in the middle. If you live north of the equator, sunlight slants toward you from the south. The farther north you are, the more extreme the angle of the sun's light. In the southern hemisphere, the opposite is true—the sun slants down from the north.

Now all you have to do is install as many big windows as possible facing the sun, and you will collect heat inside your rooms. Build your roof overhang or shades to correspond to the sun's angle, and you can cool the same inside space during hot summer days. And in all seasons, insulate above your ceiling and inside your non-glass walls to retain the inside temperature you have established.

Solar collectors carry the simple process one step further. When sunlight is available, the collectors heat water for immediate or later use. Collector panels can stand independently or be placed on your roof, angled to capture maximum heat from the sun. They come in various sizes. Passive and non-passive systems are virtually mass-produced by several manufacturers. You can learn more in a very comprehensive do-it-yourself book entitled *Build Your Own Solar Water Heater* by Stu Campbell, a Garden Way Book, 109 pages in paperback.

But now, let us get back to our 12-volt power system.

You probably have quite a list in hand at this point—all the lights, accessories, and small appliances that form the unique consumer demand of your own home. You need to begin answering some questions about the details of your installation.

● How many lights and accessories will you operate at one time?

● In what rooms will you use them?

● How many 12-volt outlets will you require in each room?

● How many on-off switches?

● Will you install an inverter? Where will your AC outlets be? How many of them?

You may want to pause here and skim through Part Two of this book. Take a look at some of the diagrams in the chapter on wiring your home. Look at the inverter installations. These details determine how effective your system will be—whether used in a wilderness home or installed as an emergency backup in a city home. The necessities of 12-volt and separate 110-120-volt AC wiring hold true in either case, and you must consider the price of wire and fixtures as you total up your costs.

Now Add It Up

Let us look at the list of items you have chosen for your system. Think about how much you will use each item in a 24-hour period. Refer to the last column in the list of accessories. These figures are the number of amps each accessory will draw from your battery storage system. When you know how many appliances you will be using, the length of time each will be in use, and the total daily amp-hours each will consume, you can determine how big your battery system needs to be.

If you simply mutliply the time in use by the amps drawn for each appliance, you will have an amp-hour figure for each appliance every day. Now you can add up the approximate total amp-hours you will consume each day. Keep in mind that you are looking for a ballpark figure, not absolute accuracy. Keep your figures loose, and you will find the process easy.

You may already have some 12-volt accessories you want to incorporate in your system. Those you want to purchase and operate will usually have amp ratings stamped on their specification plates. If there is no amp rating, you should be able to find the wattage. Divide the number of watts by volts (in this case 12), and you have the amps.

Here is an example: If you burn a 25-watt bulb for one hour, you will use 2.08 amp-hours (25 divided by 12). If the same bulb burns for just a half hour, you will use only 1.04 amp-hours, or one half of what you would draw in an hour, and so on.

By the way, if you notice amps stamped on the specification plate of 110-120-volt AC appliances you will use with a square-wave or rotary inverter, ignore them and use the wattage divided by 12 formula. That is the true amp-hour draw, because you are still using a 12-volt battery as your source of energy.

As a double-check now, compare your requirements to those in another typical situation. The following example, in chart form, represents a weekend cabin used by a family of four in the winter time. This model may not coincide exactly with your situation, but it can be compared to one person's weekly use or a city home during a temporary blackout. Have you considered most of the things this family used? Have you added others they went without?

Accessories on a typical day	Hours in use	Amps drawn	Amp-Hours required
Kitchen light	3.0	2.0	6.0
Blender			
(on AC inverter)	0.2	22.5	4.5
Water pump	1.5	9.0	13.5
Bathroom light	2.5	1.2	3.0
Bedroom lights	3.5	2.0	7.0
Living room lights (2)	6.0	4.0	24.0
Black & white TV (12V)	6.5	1.0	6.5
Stereo (12V)	4.0	1.5	6.0

Total daily amp-hour demand 70.5

You can see how simple it is. Once you have estimated your hours of use, simply multiply them by the amp draw indicated on the lists presented earlier in this chapter. In this case, the family needs 70.5 amp-hours each day of their weekend. They use the blender for the kids' juice smoothies, rather than my frozen daiquiris.

But that provides a good example of controlled AC use. Even though the blender has a high current drain in amps, it operates for only a very few minutes each day. The net cost in amp-hours is very slight for the great convenience.

You should now have a figure in mind that represents your own total amp-hour demand, based on your daily use of the system. We need a formula that compares your system use with the capacity of our old friend the battery, our power storage center.

The formula for determining use against capacity is very simple. For example, I want to operate a nine-inch 12-volt black-and-white TV with my system. I check the specifications on the set. I find they list the amperage as one. I have also checked my battery and find its capacity is 45 amp-hours. This means I can operate my TV for about 45 hours before the battery will go dead. However, I need at least 15 amp-hours capacity left in my battery to start the engine for driving the car and to recharge the battery. Therefore, realistically, I can operate the TV for only about 30 hours.

Here is another example, using wattage: I have a 12-volt travel iron with a specification plate that says it is rated at 120 watts. 120 divided by 12 equals 10. My travel iron draws 10 amps. Now if I have a 40 amp-hour capacity battery (and I don't have to think about starting the car), I can use that iron continuously for four hours before having to recharge the battery. But it will be a cold day in the wilderness before you catch me with an iron in my hand for four hours.

If your coffee pot is drawing 10 amps, and you estimate you will use it for 15 minutes, you will be using one fourth of the amp-hour rating or 2 1/2 amp-hours (1/4 of 10 = 2 1/2).

If you have decided on an inverter, here is how you plan for that. Regular appliances that run on normal 110-120-volt AC household electricity will also have a specification plate listing their wattage. The formula for figuring amps from wattage in this case is the same as for 12 volts (watts divided by volts equal amps) because a 12-volt battery is the power source. Now you have roughly the amps used by the appliance and the inverter together.

Let us say the household appliance draws 500 watts (500 divided by 12 equals 41.7). You will be pulling about 41.7 amps from your battery, plus the ten percent it takes to run the inverter. If the inverter has a 550-watt capacity, you have to add 5 1/2 amps to the 41.7, for a total of 47.2 amps being drawn from the battery in one hour. Remember that only one appliance at a time can operate on the inverter, because you will be within a hair of the inverter's capacity. More load will damage it.

Now we know the purposes you have in mind for your system, the end result. We have also done a few calculations to get a realistic estimate of the demands on your system. We are ready to start picking out the components of the system, and we can use the same principle of amp-hour use and capacity to choose your power source.

Storing Your Power

Our friend the 12-volt battery is the foundation of your system. Like the checking account for your cash, the battery system provides temporary storage of your energy, which you can withdraw as needed. In the previous chapter, you computed your energy demand in terms of daily amp-hours required. The battery you choose to supply that demand must maintain an energy balance, called its amp-hour capacity.

The amp-hour capacity of any one battery sets a strict limit on your system. It is a fixed quantity, and you cannot store more energy than the battery's storage capacity. But you can develop some flexibility and begin to increase your system's net capacity through combinations of more than one battery. Just remember that each battery operates within its own storage capacity limitations.

The rated capacity of a battery is just one of the variables. Batteries come in all shapes and sizes. They are made of different materials. There are wide differences in their cost. You can buy a new or used battery. The choices are all yours, and they are important. So we will devote this chapter to getting better acquainted with batteries.

Your battery cannot store more electrical power than its amp-hour capacity. When in use, the battery is discharging amp-hours, and the rate of discharge is the standard way of distinguishing battery capacities. The condition under which many manufacturers perform the standard rating is a 20-hour period at a temperature of about 80°F. Your ability to understand these ratings becomes very important when you want to operate a home, rather than just start your car.

You have added up the total amp-hours needed for your home in a one-day period. If you have planned a car power system, compare it to the amp-hour capacity of the battery in your car. Look for a capacity big enough to handle the job you have planned—but with enough power left over to start your car. For protection, you want at least 15 amp-hours in the battery for the surge of power needed to start your car. If your present battery cannot handle the demands of your home plus the surge power, you need a new battery, or new batteries.

When you go shopping for a battery, there are two things you will want to know in addition to amp-hour ratings. What specific guarantees cover non-vehicle use? How much deep cycling (the repetition of discharge and recharge cycles) will the battery take before its life expectancy is affected? Insist on the answers from your dealer or the manufacturer.

That advice goes for the rating itself, too. In many instances, no rating is physically visible on the battery. It may be printed only on the shipping carton or in the manufacturer's specification book. Most auto supply and battery dealers have the book and can look it up for you. Here is a typical manufacturer's list of battery identifications.

12 VOLT BATTERY TYPES

	Type	Gtee In Mo	20 Hr Rate	Surge Power
Commercial	C-24HC	24	70	380
Commercial	XH-30	24	105	525
Heavy Duty	8DH	24	205	850
Power Supply	RV-27P	12	95	530
Tractor	TC-12	18	140	650
Marine	R-24M	24	61	—
Industrial	RL4-27	36	260	—

In this example, we note that the first column identifies the usage for which the battery is best suited. All batteries, except the power supply and industrial types, are used in trucks, tractors, boats, etc. The engines in that kind of equipment run constantly, so the batteries receive a charge most of the time and are rarely deep cycled. On the other hand, the power supply and industrial types are often deep cycled by their use in standby situations where the charge is exhausted and replaced on a routine basis.

Type is the manufacturer's identification.

The *Gtee in Mo* column is the pro-rated warranty and time of service the manufacturer offers.

The *20 Hr Rate* column is the standard rate of discharge as we mentioned.

The last column, *Surge Power*, refers to the cranking performance you can depend on to start an engine at 0°F.

By calling the dealer from whom you bought the car, you can find out the capacity of the original battery. Better yet, stop by and let the dealer take a look. If you bought the car used, a previous owner could have installed a more or less powerful battery than the original. While you are there, ask the dealer to check it. See if the battery's condition is good enough to handle what you have planned.

Let us say that your battery's capacity is 45 amp-hours and that it is in good shape. If you have determined that about 30 amp-hours will be consumed in your home each day, you should have about 15 amp-hours left to start the car, according to the manufacturer's rating of 45. That *should* be adequate.

Not all battery manufacturers rate their batteries the same way. But there has been an attempt by the industry to provide a more standard rating formula, and that is what I have used here. However, these are general reference numbers only, and you should check with the manufacturers to be certain about their ratings.

In addition to the rated capacity of your battery, you need to know more about the design and materials of its basic construction. There are three basic kinds of battery construction designed for general use: normal lead-acid, industrial lead-acid, and nickel-cadmium or "nicad." But even as we work to understand these basic designs, the research and development people are busy changing the whole state of the battery art.

For most of us, the 12-volt lead-acid battery that comes from automotive manufacturers is the normal design. It consists of six electrical cells connected in series, and each cell produces two volts. The electric cell is the device that converts chemical energy into electricity, and the chemical liquid commonly used to activate it is called electrolyte.

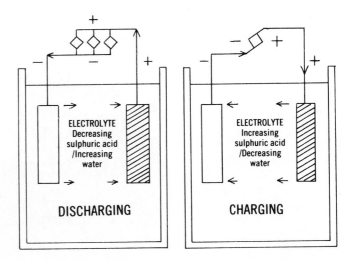

The battery runs down or loses its charge as electricity is drawn from it, usually by an accessory. (Of course you also know that simply not using a battery will also permit it to run down or lose its charge.) The electricity being drawn causes the electrolyte solution to turn to water and lead-sulfate in the cells. The amount of electricity of the charge remaining in the cell may be measured, in terms of the specific gravity of the liquid in a cell, with a hydrometer.

When you recharge a battery, the chemical reaction is reversed until the electrolyte has been restored to its original condition. This recharging process is precisely what happens after you start your automobile engine, automatically engaging the generator or alternator. Once recharged, the specific gravity reading on the hydrometer will be approximately 1.280 for each cell.

Operating on the same principle, industrial lead-acid batteries are designed for running electric forklifts, golfcarts, and the like. They can handle a hefty current drain over a sustained period of time. The capacity of an industrial battery is also rated over a period of 20 hours. As an example, if you have a battery rated at 190 amphours, you can expect to use an average of slightly more than nine amps over a constant 20-hour period.

For lower maintenance time on your system, you may want to buy an industrial battery and install it in addition to your existing normal lead-acid battery or batteries. Look in the phone book under "Batteries, Industrial."

Nicad batteries are frequently found in industrial or military back-up applications, such as for elevators in high-rise buildings or at missile-launching sites. Nicad batteries have been around for about 70 years, but the last few years have brought much improvement to their basic design. The rated capacity is different from lead-acid batteries and is based on a ten-hour constant-drain formula at 82°F. The price of a nicad is also different—it is very expensive compared to an industrial lead-acid battery.

A 12-volt nicad battery must be made up especially for you, according to your system's needs. It takes ten cells to equal a 12-volt battery because each cell produces around 1.42 volts when fully charged—as opposed to the two volts per cell produced by the lead-acid type. The advantage of the nicad battery is a claimed life expectancy of 20 to 30 years, rather than the four- to five-year maximum for the lead-acid type.

The totally different chemical reaction in a nicad battery means no acid, no sulfation, no freezing, and fewer hydrometer readings. There is also an absence of the acid fumes released by lead-acid batteries, making for a much better home installation. However, the potassium hydroxide electrolyte of the nickel-cadmium solution is highly corrosive. Nicad batteries are also much slower than the lead-acid type to recharge.

The cost ranges up to $200 per nicad cell, compared to top figures in the $50 range for lead-acid. If you can afford the initial expense, nicad may be as inexpensive as lead-acid in the long run.

Or if you can hold everything for a couple of years more, you may want to consider a redox battery. That is a *reduction-oxidization* process unveiled in 1979 by the U. S. National Aeronautics and Space Administration (NASA). Redox uses two reactant chemical fluids—chromium chloride and iron chloride—to produce chemical energy, which is then converted to electrical energy. The process pumps the fluids through a stack of cells, where the energy of each is transferred to electrodes. A special membrane prevents the fluids from combining directly with each other but permits their energy to be delivered to the electrodes. The fluids are never drained but keep recirculating through the stacks to maintain the electrical charge.

NASA scientists are testing redox models with multikilowatt capacities right now. They estimate that their new batteries will cost one fourth the price of lead-acid batteries and store far more energy. Redox batteries sound as if they will definitely belong in our powersystem just as soon as they are commercially available.

In the meantime, we need to look at some of the conditions under which our final choice of the right battery will operate. Choosing the correct battery or batteries for your needs will be determined by several factors.

1. The rated capacity of each battery.
2. The number of appliances you want to operate and how much amperage they draw.
3. How many appliances you will use at once, how often and for how long.
4. Whether or not you install auxiliary batteries, leaving one just to start the car.

Battery ratings range from about 40 amp-hours (most commonly supplied as original equipment on a car) all the way up to 400. Physical size and weight limit the maximum capacity of any battery. Although certain deep-cycling batteries are available with capacities up to 1,780 amp-hours, they are very expensive and impractical for use in a car.

In my first little house, the two rooms required little more than a few lights, TV, and stereo. I had a 45 amp-hour battery in the car and drove almost every day, 30 miles or so. I seldom had a battery problem, but then I enlarged the house and started to use my system every day, all week long. I was forced to replace my car battery about every four months. Other people told me that they tried the same thing, and they went ahead and replaced the battery each time it failed because it was still cheaper, per year, to buy new batteries than to use utility-company power. Still others had two batteries and rotated them. That gave me an idea.

My present system requires about 70 amp-hours a day, including use of the inverter from time to time and a pressurized 12-volt water system. Now I have installed auxiliary batteries both in cars and on-site at the house as well.

The first battery in the car is designed primarily for starting the car, and its life expectancy would shrink with a constant hourly drain every day. For this reason, I suggest that you use your car battery alone only if you have a vacation cabin you visit from time to time. This application assumes just a few lights and equipment that draw no more than ten amp-hours a day, and that you have the most powerful, heavy-duty battery available in your car. Even then, you may want to consider auxiliary batteries.

I highly recommend multiple-battery installations to everyone who builds the powersystem, both vacationers who use the system from time to time and two-car families who depend on it for everyday living. If you have more than one car, wire each car with auxiliary batteries. Even if your primary charging system is something other than the automotive alternator, you may find yourself depending on the cars for emergency back-up.

The advantages of an auxiliary battery system are twofold. First you gain the flexibility of being able to depend on more than one power supply. And second, you can boost the total amount of energy you store to meet your amp-hour demands.

Batteries may be connected to each other in parallel or in series. Any of the batteries described earlier in this chapter may be combined in parallel. When you parallel two or more batteries, you increase the total amp-hour capacity of your system. The voltages of all parallelled batteries must match each other, but the battery type and capacity rating need not be the same. For example, two 12-volt batteries, one with a 400 rating and the other with a 250 rating, can deliver 650 amp-hours together. Here is what parallelled batteries look like.

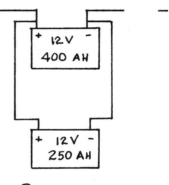

PARALLEL

You can also connect batteries in series. This combination gives you the total voltage of all batteries in the series. We will take a closer look at the details of a serial installation in Chapter 8, but for now, keep in mind that the voltage of each battery in the series can be different. For instance, you can serially connect a six-volt battery to a four-volt and a two-volt battery, and you will end up with 12 volts. The caution here is to be sure they all match in amp-hour capacity, battery type, and age. Here is a series connection.

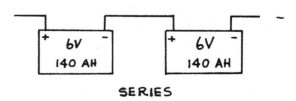

SERIES

Finally, you can connect matching batteries (same voltage, type, capacity, and age) in both series and parallel. In this combination of identical equipment, you increase both your total voltage and your total amp-hour capacity.

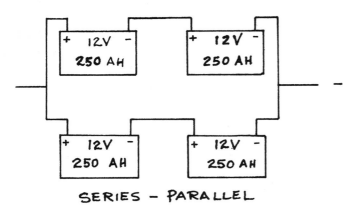

SERIES - PARALLEL

When you install an auxiliary battery system in your car, you should protect it with a dual battery-charging unit or "battery isolator" available from most auto supply stores. Here is the battery isolator in my car.

These units cost less than $20 and are easy to install. When you add the Fastcharge Overide system we will discuss shortly, your total cost is still under $40. Details for your installation are in Part Two.

Here are some final precautions to consider before we leave the discussion of battery systems in general.

Many car batteries are warranted to last four years, but nobody I know has found out how long a properly installed auxiliary battery system will last in a 12-volt home application. The average service station attendant or auto supply salesperson really has no reason to understand car batteries for any other use. My best information finally came from the San Francisco sales representative of a lead-acid battery manufacturer and a nicad representative from Los Angeles. Together with my encyclopedia, these sources have turned out to be right in my home application, and the combined information is what I have passed along in this chapter.

You have the chance to benefit now from my experience and research. The final caution is this: Do not trust just one source. Check around and compare the information you find. Experiment with it yourself—another reason to start simple and grow with your system.

Of course there is one other consideration about the batteries commercially available to us today. That is the inescapable fact that batteries, as we know them, do in fact run down. I rarely have problems with the batteries in my system. But when I do, I am forced to admit that my own negligence is the real problem. Have you ever fallen asleep and left the TV and lights on all night? I did. Luckily, I had parked the car on a slope and could start it by popping the clutch for compression. The experience was another reminder that my system, problems and all, is always my own direct responsibility.

Can you see now what is happening? You have made a list of your needs and know how much power you are going to use each day, and you have learned some of the combinations for storing exactly the power you require. You may even have in mind what you want to add to your system in the future. Notice how details start to add up. You alone control your own power-system. You do not depend on any public company.

In this chapter, you have gained some ideas about the right battery for storing the power supply to meet your demands. Remember that the batteries form your checking account of electrical power. We can turn now to the all-important deposit slip for your account—your charging system.

Charging Systems

Any electricity deposited in your battery storage system results from a translation of raw energy sources into electrical energy. The translation occurs inside an electrical generator as the external energy is converted to a rotary action, which turns your generator's flywheel. But there is more than one kind of electrical generator.

The earliest stages of the Wilderness Home Power-system depended on the obvious convenience of portable AC generators and automobile alternators. Both are still valid charging systems for us, and I will discuss them in detail. But both also depend on the combustion of gasoline, diesel, or LPG—fossil fuels—and they are becoming less convenient and more precious with every passing day.

A generator can also be started by capturing kinetic energy—the energy of motion itself—and channelling it into a rotary path. Windmills and waterwheels are two examples of how these energy sources can be applied to electrical energy generation, and we will study them in this chapter.

Kinetic energy can also be borrowed from living creatures. In many parts of the world, water buffalos, burros, or other beasts of burden are harnessed to rotary generators or automotive alternators to keep batteries charged.

Heat and light are sources of energy, too, and the subject of some very impressive current research. We will briefly examine these photovoltaic and thermopile systems, but let us begin with the obvious.

Traditional AC Generators

One of the most obvious systems for many of us in the wilderness takes advantage of the good old portable AC generator—but in a new way. Now you can operate your generator only occasionally and still have 12-volt power on a constant daily basis. The key to this variation on the system is a battery-charging device that plugs into any portable AC generator. Below is a sample of that installation.

The unit between the battery and the generator is a special kind of high-powered charger. When your generator operates, this charger takes about 15 amps of the 110-120-volt AC power and delivers up to 80 amps DC current flow to your battery system.

With the proper size generator, you can run the battery charger once a week or so, do other chores—such as washing or vacuuming—and charge your batteries at a high rate simultaneously. This system greatly prolongs the life of the generator and substantially cuts fuel and maintenance costs. It gives you abundant AC electricity for your major appliances when you need it and permits full enjoyment of the 12-volt system, too.

There are now a number of successful installations of the generator-charger-battery kind in my area. In one instance, a retired cabinetmaker and his wife moved to the mountain. The home they bought is about a mile and a half from utilities but already has a 6,000-watt generator. They do not want to run the generator every day for several reasons. They do not like the inconvenience of starting it every time they need electricity. Even heavy-

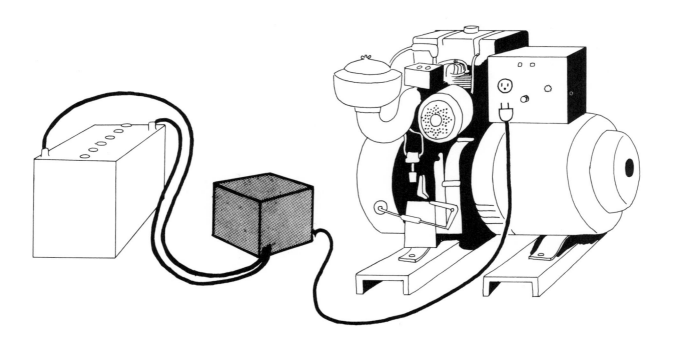

duty generators do not hold up mechanically under the strain of constant use. They do not like the generator noise, and they do not like the cost of fuel.

Their solution is a 12-volt powersystem with an auxiliary battery in their house. The charging device plugs into their generator through an extension cord just as an ordinary tool or lamp plugs in. Two wires then connect the charger to the battery.

Now, when their generator is turned on, up to 1,600 watts are used to charge the batteries, leaving 4,400 watts to run my neighbors' power saws for cabinetmaking and a regular AC washing machine for the weekly laundry, all at the same time. They plan these activities for once a week or less, and the generator, of course, is used just that often. The rest of the time, they operate on 12 volts and a 550-watt AC inverter for the light household appliances they use during the week. They have gone three weeks at a time before using their generator.

The battery charger you need is *not* similar to the kind that delivers surge power to start a car. Those chargers are relatively inexpensive because they deliver 50 to 60 amps for just a few moments to wake up a dead battery. Then they revert to their continuous rated output, usually no more than 15 to 30 amps. Battery-charger specifications can be confusing, and even many dealers do not understand the continuous rated output. In order to store maximum DC power efficiently, we need a rating of at least 80 amps *continuous* output.

Another source of confusion: some people have pointed out that their AC generators can already produce a DC battery-charging rate. A closer look reveals that they rarely produce more than 15 amps DC. At that rate, charging a high-capacity battery system will take forever.

The most readily available and economical chargers with 80 amps of continuous output generally offer another 250 to 300 amps of booster capacity to start heavy equipment connected to a depleted battery. Many units also come with timers and can be connected to automatic load sensors that turn on and shut off the generator by remote control.

Check specifications carefully before buying your charger. You should expect to pay about $200 for a good 80-amp continuous-output model with booster capability and a timer. You may be quoted a price as high as $300, but that will include 100 amps of continuous output and more than 300 amps on the booster—features you rarely need for a home charging system.

Two charger models I have tested and used successfully during the past two years are the Dayton 3Z632 and the Solar 400. Both have ratings of 80 amps continuous duty and 250 amps momentary booster. They are also equipped with timers, charging rate selectors, and ammeters to monitor the DC output. If you cannot find these models at your dealer, call or write to these manufacturers for more information.

- Dayton Distributors: W. W. Grainger, Inc. 5959 West Howard Street, Chicago, IL 60648 (312) 647-8900

- Solar, A Division of Century Mfg. Co. 9235 Penn Avenue, Minneapolis, MN 55431 (612) 884-3211

Let us consider the generator itself. If your property is not already equipped with one, you have the chance now to make a careful choice. You must consider the following points.

- Local regulations
- Kilowatt capacity
- Fuel
- Construction and optional features
- Installation and maintenance.

Checking local regulations should be your very first step. If you live in an area subject to city or county planning departments, you may find prohibitions against the use or storage of certain fuels, including either gasoline or propane, which may power a generator. As populations become more dense, legitimate noise levels may well be restricted, too. But you can often mitigate the problems to acceptable standards through your choice of relatively safe, quiet equipment. Then take care to house the equipment in a well-ventilated, properly exhausted, soundproof enclosure.

The kilowatt capacity to meet your requirements is another primary consideration. As you list your own electrical demands and begin to explore the different sizes of generators available, you need to keep in mind two simple formulas. The first is a slightly different formula than the one we learned earlier. It is used to read appliance specification plates: *Watts = Volts x Amps*. (An example, 117 volts x 43.0 amps = approximately 5,000 watts.) Then as you talk to dealers, you need to know about *KW*, the abbreviation for kilowatts or thousands of watts (5,000 watts = 5 KW). If the dealer asks what KW unit you want, and you have figured on 5,000 watts, your answer will be a 5-KW generator.

With your local regulations and these two formulas in mind, here are four guidelines to help you choose the proper size AC generator.

1. Add the total wattage of all appliances, tools, and lights you have listed. This list goes beyond the 12-volt demand you established in Chapter 2. Now we are talking about 110–120-volt AC equipment, like the washing machine, shop tools, and deep-well water pump, which you want to use on a controlled periodic basis—and we are talking watts, not amps. Remember, too, that your battery charger itself is an appliance that draws wattage from the AC generator. Add at least 1,600 watts for the charger to the total AC wattage demand on your generator.

2. Be aware of the difference between *starting* watts and *running* watts. Most heavy-duty appliances depend on a capacitor or split-phase motor, which draws up to four times the rated wattage for a few seconds in order to start its operation. Once started, the motor reverts to its running wattage, but your generator must be capable of providing the starting watts for those few seconds.

The following appliances use a capacitor to start: compressors, large fans, business machines, refrigerators, air conditioners, pumps, mechanical doors, and some table saws, drill presses, or lathes. Split-phase motors will be found where moderate starting torque (the force that begins rotation in the motor) is required, as in blowers, grinders, centrifugal pumps, gear motors, belt-driven fans, and certain other tools. If you cannot identify the kind of motor in a piece of equipment, check with your dealer or the manufacturer.

In calculating your total demand, you must figure the starting watts. Although you may leave all your appliances running, the motor in your washing machine, like several other motors, will work through its cycles ("wash" and "rinse") a number of times as it runs. With each new cycle, the motor draws starting watts again.

Here is a chart you can use to check starting watts and running watts, based on the size of the motor.

1725 RPM* Motor Size	Capacitor Starting Watts	Split-Phase Starting Watts	Running Watts
1/8 HP	400	800	225
1/6	600	1,000	275
1/4	800	1,500	400
1/3	1,200	2,000	500
1/2	1,500	2,500	700
3/4	2,100	3,500	950
1.0	2,500	—**	1,200
1.5	3,000	—	1,600
2.0	3,500	—	2,200
3.0	4,200	—	3,200
5.0	7,100	—	4,700

* If the motor speed is 3,450 RPM, increase the starting watts by 15 percent. For particularly hard starting or high inertia motors, such as compressors and blowers, add another 25 percent to the starting watts.
** Split-phase motors become very inefficient in larger sizes and are rarely manufactured.

3. The rated horsepower (HP) of your generator is based on operation at sea level, fueled by gasoline or diesel. Once you have listed your total wattage demands, including the allowance for starting watts, you must correct your figure for actual operating conditions.

For each 1,000 feet of altitude above sea level, your generator operates about 3 1/2 percent less efficiently. As an example, if your mountain is 2,000 feet above sea level, and your rough total demand is 5,000 watts, you must increase your total wattage by about seven percent, making it 5,350 watts. Since generators are usually rated in 500-watt increments, you would ask for a 5.5-KW rating or 5,500 watts.

For LPG or propane fuel, you need to derate the total output by another ten percent. In the example just used, add another 535 watts to the 5,350 for a total demand of 5,885 watts. You need 6,000 watts or a 6-KW generator.

4. The final guideline is based on practical experience by my neighbors and in my own household. As technology continues to tempt us with new products and experiments to try, we want to anticipate future demand. Our rule of thumb calls for adding 25 percent more capacity than our immediate needs.

To recap, assume that you plan to run a water pump once a week to fill your holding tank. The pump is rated at one-half HP with a capacitor motor operating at sea level. From the chart, you can see that you need 1,500 starting watts. At the same time, you want to put one week's clothes in the washer, which has a half-HP split-phase motor. So you add another 2,500 watts. Of course you want to charge your battery system while the generator operates, so add 1,600 more watts. Your subtotal is 5,600 watts. Now you add another 25 percent for future expansion, which means you really need 7,000 watts, a 7-KW generator.

A 7-KW generator is expensive. Is there any way to reduce the demand, to economize on the initial outlay? Only one logical way. Sacrifice some of your allowance for future expansion to about 10 percent. You can get an output of 6 KW for your immediate needs and still have a slight margin for the unforeseen.

On the other hand, if you are thinking of much lighter duty for your generator, a short operation once a week or less, you may want to look at portable AC *alternators*. These machines do the same thing as the generators, but they are not built for continuous heavy-duty use. You pay considerably less, and you can still get most of the generator's features and options. Two cautions: many alternator models cannot handle high motor-starting watts, and full wattage output may depend on 220-240 volts AC, which is double what most of us use. Check the specifications carefully.

Getting back to the list of basic considerations for your choice of either a generator or alternator, think about the kind of fuel you want to feed your charging system. Gasoline has been the standard for many years, but as the price climbs and the availability dwindles, some alternatives become more interesting.

Generators that burn diesel fuel are more economical than those that use gasoline. In addition to lower cost per gallon, diesel units can be more efficient, using 1/2 to 3/4 gallons per hour, rather than 1 to 1 1/2 gallons of gasoline. The basic design of diesel engines makes maintenance easier, because it avoids the conventional ignition system. The initial cost of diesel equipment, however, can be up to $1,000 more than its gasoline-fueled counterpart.

Conversions to burn LPG, propane, or natural gas can be very economical, too. These fuels cost even less than diesel fuel and burn much cleaner, resulting in still lower maintenance. If natural gas is available from a nearby public utility, there will be very few problems, but be sure you check local legal restrictions carefully before storing any fuel on your property.

Most generators can be converted from gasoline operation by replacing the carburetor with one modified to burn the fuel you specify. Conversion kits are available from many Honda dealers for about $200. You may also find them by writing to Beam Manufacturing Co., 3040 Rosslyn Street, Los Angeles, CA 90065.

Just one note about your machine's construction: generators were traditionally built with cast-iron blocks, but some manufacturers have lately tried lightweight aluminum. I have heard some stories of premature engine failure from people who bought the aluminum-block units, so you would be well advised to question the materials when you buy.

As you look at generators or alternators at the dealer's store, you will want to know about several optional features with which the various models may or may not be equipped. Depending on your operating demands and how much you want to pay, here are some features to think about.

- Manual, 12-volt, or automatic demand. The automatic control signals a unit to start and stop through the activity of any appliance on the AC circuit, or it can be set to cut in when another power source fails. This is an important feature for city dwellers who want a unit as back-up to utility company power.

- Continuous-duty rating. This feature assures efficient operation at the fully loaded rate on a constant basis.

- Full wattage output. Protect yourself from generators that require distribution of the 110-120 volts on two AC circuits, which means you get only half the rated capacity on one circuit.

- High motor starting. Special windings deliver extra power for your required starting watts.

- Inherent voltage regulation. The voltage output adjusts automatically to match your required load, without affecting your special external battery charger's rate.

- Two-way engine starting. Manual or hand-cranking start serves as an alternative to the usual 12-volt or automatic demand.

- Idler control. You can reduce fuel consumption and

engine maintenance by this automatic RPM adjustment to match the load. One caution here: a low-wattage motor alone may not cause the generator to react sufficiently. To correct, just add to the load, such as turning on a 25-watt lamp.

At this point, you may be getting a fairly specific idea of the AC generator you need to charge your 12-volt batteries as it meets various other controlled needs in your household. But there are two other basic considerations to bear in mind before we hurry off to our nearest dealer—installation and maintenance.

When you contemplate installing an AC generator and probably some AC house wiring, you are thinking about extremely high voltage and the potential for serious electrical shocks. If you are not now a qualified electrician, please consult someone who is—especially if public utility lines are involved anywhere in your installation.

In the city, your generator may share the AC wiring served by public utilities. But if it is installed improperly and you are using the lines during a power failure, you could be responsible for bringing serious harm to utility people working on the lines.

Once your machine is installed, the availability of parts and service plays a critical role in ongoing maintenance. Be sure that the parts needed do not have to come from too far away. And remember that each time you cannot fix a problem yourself, a service call will cost $50 or more—and substantially more if you live in the wilderness.

Okay, if an AC generator-charger-battery system makes sense for your home, you should be ready to go shopping. For a high-performance AC generator, gasoline-powered, started manually or on 12 volts, with a 5-KW output, expect to pay around $2,000. If you want an output up to 7 KW, $2,600 is a reasonable price. On the lower end, a 2-KW or 2.5-KW unit will cost less than $1,000.

For lightweight AC alternators, gasoline-powered, manual starting and with no options, the 2-KW or 2.5-KW models will cost about $700. The largest alternators I have seen are 4 KW for a little over $1,000. The smallest made is Honda's .5-KW machine for about $300.

Honda gets very high marks for quiet, reliable equipment and good service. Dayton builds a good unit, as do Kohler and Onan, who also offer diesel-powered units.

Kohler and Onan both manufacture alternators and generators used widely in RVs, because the engines are compact, quiet, and enclosed to permit routine maintenance from just one side of the installation.

Here is the complete list of reliable generator manufacturers. I have used A when they also build alternators, D when diesel equipment is available. Write to them or call when you need information about special applications or to find your nearest dealer.

- Cummins-Sandstand, Inc. (D)
  1000 Fifth Street, Columbus, IN 47201
  (812) 372-7211
- Dayton Electric Manufacturing Co. (A)
  5959 Howard Street, Dept. TR, Chicago, IL 60648
  (312) 647-0124
- E. R. Ferguson Co. (D)
  142 Otter Street, P. O. Box 222, Bristol, PA 19007
  (215) 788-1774
- Homelite, Division of Textron (A)
  15501 Carrowinds Blvd., Charlotte, NC 28217
  (704) 588-3200
- Honda Motor Company
  (Widely based distribution. Check phone book.)
- Kato Light Corp. (A)
  3201 Third Avenue, Makato, MN 56001
  (507) 625-7973
- Kohler Co. (D)
  44 High Street, Milwaukee, WI 53004
  (414) 457-4441
- Merican Curtis Inc. (D, 3KW and up)
  1717 Rittenhouse Square, Philadelphia, PA 19103
  (215) 985-1313
- Onan Co. (D)
  1400 Seventy-third Avenue NE, Minneapolis, MN 55432
  (612) 574-5000
- Pincor Products (D)
  5841 West Dickens Avenue, Chicago, IL 60639
  (312) 237-4100

All this detail about portable AC generators is not intended to suggest that you use them to supply all your needs. I am talking about them as one important possible element in a combined system that depends usually on the 12-volt power we store in batteries. With a charger connected between your battery and your AC generator or alternator, you gain freedom from the traditional total dependence on these noisy and usually expensive machines.

The cost of fuel continues to soar faster than the rate of general economic inflation. If you choose one day a week to run your AC generator, you can pump water into your holding tank, do the laundry and vacuum cleaning, and charge your batteries all within the same five- to six-hour period. Your cost will be under $10 a week, a tremendous saving compared to the cost without a 12-volt powersystem. The folks who live up the hill, who run their generator for up to six hours every day to keep all their lights and appliances on 110–120-volt AC power, spend more than $50 a week, plus the cost for more frequent maintenance.

So the controlled use of the good old generator is one bona fide alternative, but think about the future cost and availability of fuel. Do not decide on the generator until we talk about some other alternatives.

Automotive Alternators

In its earliest stages of development, the Wilderness Home Powersystem did not depend on portable AC generators, or on wind, water, sunlight, or heat alternatives—which I had dismissed as too technical or too exotic for me. My primary charging system was under the hood of the family car. That one battery ran just a few lights and accessories in my small house and was automatically charged when I drove down the mountain every day. I will cover some of the details now, but first take a look at this bird's-eye view under the hood.

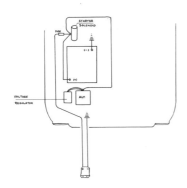

Simple, is it not? Driving in the daytime puts a charge on the existing battery, which stores electricity to be used at home that night. Of course, we have to keep an eye on our energy use to be sure there is enough power in reserve to start the car.

It works, and I proved it for a couple of months. But this simple system soon becomes a little too simple. I was quickly spoiled by the first conveniences and began to make greater demands on my system. I learned to expand on the basics, and I found out how to modify them, overcoming some built-in problems along the way. Looking back now, I realize that the modifications are also simple.

As we build on the basic system, let me urge you again to install an auxiliary battery. That is the most convenient and reliable solution to storing the continuous power you will find yourself needing. Your imagination sets the only limits on where to put the auxiliary batteries. Here are some sample installations.

In your car, recognize the importance of ventilation and a sturdy box or secure brackets (which may not show clearly in these photographs) to prevent the batteries from tipping over and spilling corrosive acid.

If you have two cars, wire up the second one as well. Then when one person leaves, the other plugs in the second car during the day. You will naturally use less electricity during the day than at night, but remember to charge the second car's battery with an occasional drive or by idling the engine.

If you have only one car and it is gone during the day, you can locate an auxiliary battery on-site at the house, as well as in the car. Details for a good on-site installation form part of the instructions in Part Two.

Auxiliary battery installations in all cars and on-site at my house have given my total system great flexibility and reliability. And we can still use an automotive alternator to charge all the batteries, usually with no more than normal daily driving. The car's engine does not have to be running to operate my home, and battery isolation devices keep the auxiliary batteries disconnected from the first battery, so there is always a charged battery to start the car.

To build this simple system yourself, you need to understand just two basic principles. The first is how your automotive alternator works. The second is the difference between *voltage-regulated* charging and *current-regulated* charging. Let us take them one at a time.

Understanding the alternator depends in part on the same amp-hour formula you used to add up the demands on your powersystem, this time in reverse. If your alternator will produce 30 amps output at capacity, it can deliver 30 amps to your battery under constant charging conditions. If the alternator is rated at 55 amps output, it can deliver 55, and so on.

The alternator produces a lot of 12-volt energy in a short period of time. You can use the energy right away or store it for later use. You will want to be sure your alternator has the output to keep up with your energy withdrawals. I will not present a full technical description of the alternator, but I do want to give you a few basics. Here are two views of a typical automotive alternator.

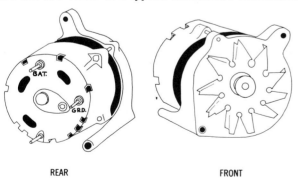

REAR   FRONT

The pulley action in this unit produces alternating current (AC) when the alternator is operating at certain revolutions per minute (RPM). It can be driven by your car engine or another external drive force. Simple diodes are used internally to allow the current generated to flow in only one direction, thereby converting it to direct current (DC) to feed the battery. At 400 to 900 RPM, depending on the manufacturer of the device and the pulley size ratio, the alternator begins to produce amperage. The faster the RPM, the more amperage generated, until the maximum output is reached.

The chart below is an example of the Ford alternator (GPD side terminal) specifications and ratings, 1972-76 models.

Amps @ 15V	Watts @ 15V	Field Current (amps @ 12V)	Cut-in speed (eng. RPM)	Highest rated output (eng. RPM)
65 (1972 only)	975	2.9	360	1640
70	1050	2.9	725	5000
90	1350	2.9	875	5000

You can see that to achieve the maximum charge rate listed in the first column, the engine must run at the RPM listed in the last column. The higher the rated output of your alternator, the less time you must operate your car's engine to charge your batteries fully, except for the limits set by the voltage regulator.

Different kinds of alternators are made for different makes of cars, but all operate basically the same way. Generally they are available at a rated output of 30 to 105 amps. Alternators of 30- to 70-amp output are available from auto supply stores and car dealers. Higher ratings are made for trucks and tractors, and some of these are adaptable to passenger cars or light trucks. The higher the rated output, the faster a battery will charge—once the voltage regulator is bypassed.

To find the rated output of your alternator, check with the manufacturer, or look in your car's operating manual if you still have original equipment. The surest way to check your alternator's output is with an ammeter, as shown here.

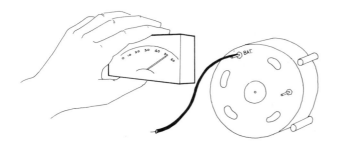

This ammeter reads current by measuring the strength of the magnetic field around the wire. There are no connections to hook up. Just hold it next to the "Bat." output wire from the alternator with the engine running, and it will indicate the amount of amperage generated. You may be able to borrow an ammeter from your mechanic, but they are also available at auto parts dealers. Many different types are made, and some cost as little as $10.

For the most accurate check with your ammeter, first turn on all your car's lights for about 15 minutes. Then with your lights still on, start the engine. Pick up your ammeter, and take your reading at once with the engine at 2,000 RPM or more. Your lights will have depleted the battery somewhat, so your alternator must begin its output closer to its maximum rating. Remember to turn off your lights after the test.

If you decide to buy an alternator with higher amperage, I suggest you check *Petersen's Big Book of Auto Repair*. It contains good illustrated descriptions of all American-made alternators and their ratings, as well as how to test and repair them. I have reprinted Petersen's alternator checklist in Part Two.

Unless you buy a new car, the alternator you purchase is probably rebuilt. Its cost at an auto parts store ranges from about $50 to $130. Used alternators, on the other hand, are available from auto wreckers at much lower cost. If there is no way to test a used alternator before you buy it, be sure you can return it later if it does not operate.

Now to the second basic principle. In order for your

batteries to get the maximum charge from the alternator in your car, you need to bypass the voltage regulator on a carefully controlled basis. Your voltage regulator plays a vital role in protecting the regular car battery from being overcharged and damaged during the engine's normal operation. When it senses the battery filling with energy, the regulator reduces the charging voltage generated by the alternator. The battery actually determines its own charging rate, which will taper off to nothing if and when a full charge is reached. With this protection, there is no limit to the length of time a charger can be left connected to the battery.

Your car's engine was built to move you about, not just to turn the alternator and charge the batteries. The voltage regulator limits the charging current to your battery, according to the voltage levels of the battery—maintaining just the level necessary to keep your battery in top condition. It does so by actually decreasing the current through the charge cycle. For simplicity, I have divided the cycle into four stages. With the voltage regulator, the first quarter of the total charging cycle occurs very rapidly when you first turn on the engine. The second and third quarters of the cycle take progressively longer to accomplish. And the final stage for topping off the charge may take up to 30 times as long as the first quarter of the charge did.

If we bypass the voltage regulator, we can keep the alternator's output voltage rising sufficiently to speed up the last three quarters of the charging cycle for as long as the alternator is allowed to operate. Rather than a constant-*voltage* charging system, we have a constant-*current* charging system.

Constant-current charging will not work if you have a single battery for both your home and your car, because you eliminate the protection for your starting battery, but it is a perfect complement to an auxiliary battery system. By flipping a switch on your dashboard after you start your car, you can bypass the voltage regulator while driving. The car operates on the energy stored in your starting battery. Meanwhile, constant-current charging permits your alternator to charge your auxiliary batteries at the most rapid rate possible while you drive, tapping all of your alternator's capacity. Shortly before you reach your destination, switch off the bypass. Voltage regulation is restored, and your alternator begins to replace amps used from the starting battery to run your car.

Obviously, the constant-current charging system is

designed to store a lot of 12-volt power fast. The system would be wasted if used with typical light-duty automotive batteries, but it serves well with deep-cycling industrial batteries, those with a capacity of 120 amp-hours or more. An industrial-capacity 12-volt battery of 250 amp-hours meets my needs, and I paid about $250 for it. So the initial investment in an auxiliary battery may be a little steep, but the fast-charge capacity permits us to get a lot of stationary "mileage" out of the equipment.

In practice, the economics of constant-current charging make good sense. For example, an 85-amp alternator in my car produces a constant 80 amps when I switch on the current-regulated system. I am allowing the voltage to increase as much as necessary to hold the constant 80-amp output. Now my 250 amp-hours of charge are delivered in just over three hours instead of 25 hours under voltage regulation, and I burn just 1 1/2 extra gallons of gasoline instead of five.

One important caution in constant-current charging:

Do not try to operate the 12-volt appliances in your home powersystem while charging the batteries. With no voltage regulation, your charging current maintains a constant 80 amps during its rapid cycle. That can translate into as much as 20 volts output from your auxiliary batteries in the final phases of the cycle. In other words, it could spell disaster for your 12-volt appliances. Be sure your house is unplugged during charging, or that the charging system stops when the powersystem is plugged in.

You will find instructions in Part Two for installing the constant-current charging system. You may also be interested in the simplified installation of a constant-current charging kit. My engineer friend Clyde Davis has helped to develop the only commercially available systems we know, the Custom Powercharger and the Fastcharge Overide System. Here is a look at the Powercharger, and full descriptions of it are in the catalog for which there is an order form at the end of this book.

In my combination of the car's alternator with a fast-charging capability and the 12-volt powersystem, my wilderness home has uninterrupted power, but here are a few other tips I can offer from experience.

- Your alternator can deliver its maximum output as you drive at speeds as low as 10 mph.

- Night driving delivers a lower charge rate to your auxiliary battery because your car lights, radio, etc., must draw some amperage directly from the alternator. Use of the constant-current system is not recommended for night driving.

- You must still drive enough during the week to keep up with your energy needs when you use the auxiliary battery system in the car to run your home.

Take a look now at these samples of driving time and charge rates based on the use of a constant-current system.

| | Charge Rates for Alternator Outputs | | | |
Driving Time	30 Amp	50 Amp	60 Amp	70 Amp
15 minutes	7.5	12.5	15.0	17.5
30 minutes	15.0	25.0	30.0	35.0
60 minutes	30.0	50.0	60.0	70.0
90 minutes	45.0	75.0	90.0	105.0
120 minutes	60.0	100.0	120.0	140.0

These charge rates must be adjusted by the condition and efficiency of your particular battery. Think of it as a service charge on your energy checking account that may run as high as 25 percent. Even if your alternator charges at a 100-amp-hour rate, your battery may deliver only 75 amp-hours into the house.

In the case of an auxiliary battery on-site at your house, you should install a heavy-duty battery of smaller amp-hour capacity than the auxiliary system in your car. Then when you arrive home and connect the house to the car, you immediately replace most of the amp-hours in the smaller battery. The wire sizes in this connection are critical, though, because of the rapid surge of capacity amps. Check Chapter 6 regarding the correct wire to install.

When the on-site battery's charge becomes low, plug in the car and run the engine for the length of time it takes to restore the full charge. If your constant-current charging system is engaged, you cannot operate any 12-volt components or appliances while charging batteries this way. Refer to the chart for the correct time.

That about covers the automotive alternator choices. But what if you do not drive enough and do not want to plug in the car with the engine running just to charge batteries? What if you simply do not have a car? Here is a potential solution.

You do not have to drive an automobile to use an automotive alternator for charging. In my prototype Wilderness Home Powercharger last year, I teamed an alternator with a lawnmower engine for an externally driven charging system. During the three months I used it, I kept a diary on its operation. I ran the engine for a total of 41 1/2 hours to provide myself with 555 hours of available electricity, using an average of 1.85 amps per hour. Now a friend, who lives in Michigan, has taken my first experiment to the point where it provides dependable primary charging for his powersystem.

Both alternators and traditional generators depend on motors that run on fossil fuels. As those fuels become increasingly rare, many of my friends and neighbors in the wilderness have begun turning to even older sources of energy, whose potential practical applications are still rather novel. I mean wind, water, and heat power, converted to charging systems for 12-volt batteries. These are the same charging systems I wrote off as too technical or too exotic just a little over a year ago. They look much simpler and more practical all the time.

Wind Plants

Although the wind's energy can be captured by means that are anything but exotic, there are some important technical considerations to make a system work. They fall into two main categories: site evaluation and suitable equipment.

You may not realize that those picturesque drawings and photos of simple windmills in rural areas do not reflect all that is really needed to produce electrical energy from the wind. A closer look at the actual mechanics would reveal some careful planning beneath the picturesque surface. Now as I work at my little store in Laytonville, I keep meeting people who are deter-

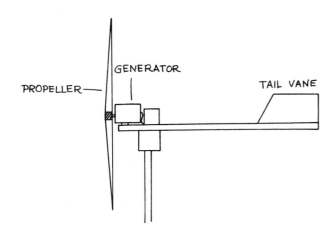

PROPELLER — GENERATOR TAIL VANE

combined with a pair of binoculars in hand back on the ground. You have probably seen these anemometers at the airport or other weather stations. They are little wheels that turn as the wind is caught by cups fixed to the ends of the spokes. You can buy a mechanical anemometer through Sencenbaugh Wind Electric, and it should not cost more than $175. You can also rent them.

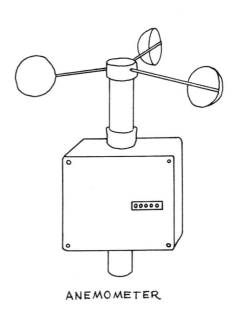

ANEMOMETER

mined to make the wind work for them. Some of them have enjoyed great success, but some of them keep running into disappointments. They say they "just know there is plenty of wind" but fail to capture enough of it. Or they go to the other extreme, unleashing a power source that, in real waste, far exceeds their expectations.

One of the more successful stories is Jim Sencenbaugh, now President of Sencenbaugh Wind Electric in Palo Alto, California. Jim's *Catalog 1078* provides far more than a list of parts, describing site evaluation and the many choices to be made in a safe and realistic decision to harness wind energy. I am grateful to Jim for many of the practical details that follow. You can get his full 48-page catalog by sending $2 to Sencenbaugh Wind Electric, P. O. Box 11174, Palo Alto, CA 94306.

Both the problems of too little and too much wind can be solved with the first step in our technical approach, site evaluation. Before you reach into the sky with any wind plant, you need to make a careful assessment of your winds' usual character (your *prevalent* winds) and make a reasonable allowance for those unusual gusts of high energy (your *storm* winds) that will almost certainly sweep your site at some point.

What is the wind speed on your site? Generally you need an average monthly speed of eight to ten miles per hour or more for a charging system. To determine your average wind speed, you will erect a temporary tower or a sturdy pole with guy lines, 30 to 50 feet high. At the top, you will place your wind-measuring device. For most of us, a mechanical anemometer is the device,

As your anemometer turns at the top of your temporary tower, it measures and registers the distance a column of air moves in a given period of time. The wind speed is then calculated, according to the manufacturer's formula, over a period of 30 days.

This monthly average wind speed is only a base figure on which to begin your calculations. What you really need is a long-term average for every month in a whole year. But you do not want to sit there looking through your binoculars for the whole year every time you pick a new site to test.

Government weather stations in your area can provide average monthly figures, comparable to your own, but for the whole year. Contact the weather stations closest to your site, preferably at least three of them, and arrange to compare your records to theirs. Starting with your base month, compare their averages to yours. You want to find a ratio between your figure and each station's figure for exactly the same month. Then you can find another figure that is the ratio between your monthly figure and the average of all their figures together.

For instance, if your own winds during the sample month are 1.10, 1.40, and 1.25 times as great as each of the three closest stations for the same month, then you know that your wind's average speed is 1.25 more than their composite average speed. If those three government reports average out to 10 miles per hour, the average speed on your site is 11.25 miles per hour.

Of course, your average wind speed for just one month is not the whole story. The wind speed in August, usually the least windy month, will differ greatly from December, when storms are frequent. You can use your ratio, 1.25 in the example, from the one month sampled to estimate your site's wind speed in any other month. Just find the composite averages for the government stations for each month of the year, and multiply them by your ratio factor.

You do not want an annual figure, just 12 monthly approximations, because your main interest is in electricity. In December, your electrical demand may be for evening lighting from 5:00 to 11:00 p.m. But in August, your demand will only be from 9:00 to 11:00 p.m. Mother Nature seems to have thought of everything for us.

But Mother Nature must also assume responsibility for truly violent weather. Your local weather stations can also supply information on storm winds, which are the speeds of wind gusts (in terms of the *fastest mile*) for each year since they began keeping records. As an example, assume the fastest mile in the last 25 years at these stations was 70 miles per hour. With the 1.25 adjustment ratio for your site, your fastest gust in the same time can be estimated by the following formula. The figure 1.33 is the gust constant used by weather stations to complete the formula.

$$70 \times 1.25 \times 1.33 = 116 \text{ mph}$$

Your chances of having another gust of the same strength in the next 25 years are just about 100 percent. Your wind generator system must be able to withstand that force.

As you conduct your site evaluation, remember that wind activity occurs in physical space. With this principle in mind, you can avoid some of the most obvious interference with maximum winds just by looking around carefully when you select the site for a wind plant. Here are some examples.

A.

A sloping roof deflects wind currents from their normal path. But if your prevalent winds are in line with the roof's ridge, your tower needs to be on the same line. The solution is to place your wind plant well away from the house, at a point where the deflected wind currents will return to their normal path.

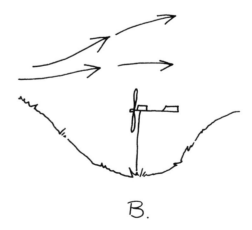

B.

Frequently you can measure good, strong winds running down the slope of a hill. A valley appears to be a good place to trap those winds, but you will also discover that downhill winds only come from one direction. Depending on seasonal conditions, the top of the same hill sits in the path of several more winds than the bottom. Your wind plant deserves its chance at all of them.

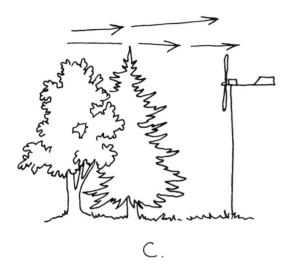

C.

The air currents directly over a tree or cluster of trees are very interesting. Because trees are flexible obstructions, changing shape in the wind, they shift the winds and interact with other obstructions as well—including wind plants. From 15 to 20 feet above the trees, the path of winds may proceed along steady lines, but below that height, winds are turbulent and change directions rapidly. Assuming you do not want to chop down these trees on your property, you have two choices. Either build your tower high enough to stay out of the turbulence, or select a site well away from the trees.

D.

Building your tower right in front of the trees is no solution either. The turbulence causes the wind currents to begin lifting away from obstructions even before they reach them. Although your tower may be in the face of prevalent winds most of the time, it becomes part of the pattern of turbulence, and the winds you want are effectively trying to escape from that pattern.

With these examples of interference in mind, you should see already that you cannot erect your wind plant and tower on top of your house, or any other roof. Assuming that you placed it high enough to avoid interference, you would still be inviting a couple of other problems. One is the transmission of noise and vibrations into your structure whenever the big propeller turns. The other is the danger of losing your house, as well as the wind plant, in high storm winds.

This quick summary of some basic principles in site evaluation cannot stand apart from knowing the basic components of the wind plant and what they do. A sturdy tower is a must. Other parts are the rotating blades of the propeller, a low-speed DC generator or alternator, the tail vane or rudder, and usually some form of lightning protection.

The tower must stand at least 20 feet higher than the nearest obstruction and high enough so that the diameter of the circle through which the propeller turns does not encounter any interference. For most of us, the tower will be 40 to 50 feet high, and it must be strong enough to withstand the fastest gusts of storm winds we have already computed. Tower construction can be one of three kinds.

The first kind is the homemade wooden model, the least expensive but the trickiest to build from an engineering standpoint. Unless you are a seasoned old-time carpenter who really knows the integral strength of wood grains and joints, you will save yourself much heartache by dismissing this category now.

The second construction option is a guyed metal tower strengthened by cables radiating out from a point high on the shaft and secured to the ground. This construction is fairly economical in terms of materials, but it can be very costly in terms of available space. For sufficient strength, guy wires should be anchored at a horizontal distance from the base of the pole that is equal to 80 percent of the tower's height. If the pole is 50 feet tall, each of at least three wires needs 40 feet of horizontal space—an area 80 feet in diameter all around the base of the pole with no trees or structures inside.

The best wind towers are built on a welded triangular design of high-strength, galvanized steel tubing. The sections of the tower should be ten feet long, inserted one into the next for rapid, easy installation. The materials for this construction alternative are the most expensive, but the resulting tower is the strongest and requires the least maintenance once it is in place. One of the most established and reputable firms supplying most wind-plant manufacturers is Rohn Towers. You can call or write to them directly for details via Unarco-Rohn, 6718 West Plank Road, P. O. Box 2000, Sioux City, IA 51102. Telephone (712) 252-1821.

You can expect to spend anywhere from $600 up for a good self-supporting tower.

At the top of the Rohn Tower or any other good self-supporting tower are the working parts of the wind plant. The key is the axle on which the propeller turns, the axle itself free to rotate laterally as the tail vane finds the wind direction. Then all this motion is translated into electric power by the generator. This connection must be secure and mechanically sound in any wind speed.

Propellers capture the wind's linear energy and convert it to the rotary action necessary to drive the generator's flywheel. But as the blades of the propeller turn, each blade creates disturbance for the blade following it. The more blades on a propeller, the greater the disturbance and the lower the overall performance. Four or more blades work fine for water pumping, where the design calls for low speed and high torque. Two blades produce gyroscopic instability, competing with the tail vane to orient the whole unit into the wind. However, a two-bladed prop works very well in a properly designed small wind plant. For a larger plant, the ideal number of blades is three.

Each blade represents the radius of a circle through which the rotor turns. The greater the diameter (two times the length of one blade) of the circle, the more wind the rotor will capture. The main factor in blade design is that the amount of energy captured by the rotor depends upon the disc area, or outside diameter of air space, intercepted by the rotor. This relationship between propeller diameter and output is proportional to the square of the propeller diameter. In other words, if you double the size of your propeller, the power output will increase four times. But here is where you must balance the maximum output from prevalent winds against the safety required in storm winds. Your whole

wind plant must be strong enough to withstand a storm's increased load, and the larger your rotor's diameter, the more of that load you will capture. A rotor's diameter of 10 or 12 feet will be easier to handle structurally than a diameter of 22 to 30 feet.

The actual power available from the wind is the mathematical cube of the wind speed. Put in simpler terms, if a ten-mile-per-hour wind speed is doubled, you will have eight times the power available to the rotor. But in practice an open-air wind plant only recovers about 59 percent of the kinetic energy.

You can see why the relationship between wind speed and rotor diameter is important, especially if you have ideas about building your own plant. The job is not easy and takes considerable engineering to avoid danger. I am told that there is nothing quite like a large propeller or generator flying through the air after having been torn from its tower in storm winds. I believe it.

The tail vane acts as a rudder, catching wind currents in a vertical plane to orient the rotor into the prevailing wind. Like most of these components, it is engineered and chosen carefully to correspond to the generator and rotor size.

Your wind *generator* is identified by a wind-speed rating, based on its maximum potential output under ideal conditions. The published wind-speed rating of any generator depends on many physical variables and can be influenced to some degree by salesmanship, testing procedures, and applications. If the rating is to have any value to you, insist that a dealer clarify it in terms of your site, according to your wind-speed data, at your altitude. Most commonly, ratings are for the air density at sea level. Higher altitudes, with lower air density, require a higher wind speed to obtain the same electrical output. Temperature will also have an effect on the output.

Altitude, height.	DRA (60° F.)
sea level	1.000
2,500 ft.	0.912
5,000 ft.	0.832
7,500 ft.	0.756
10,000 ft.	0.687

This chart should help you get some idea of the real output for a wind plant rated at your site's altitude. The density ratio altitude (DRA) is based on a temperature of 60° F., which provides a typical norm without getting into too many formulas. From the chart, take the DRA figure closest to your own site's altitude, and multiply it by the rated output of your generator, according to the manufacturer's specifications. As an example, a unit that is claimed to deliver 500 watts at a given wind speed at sea level will actually deliver about 416 watts at 5,000 feet. (500 W x 0.832 = 416 W.)

Most well-designed wind generators are equipped with voltage regulators to protect your batteries from overcharging and a blocking diode to prevent power from returning to the generator when there is no wind.

Lightning protection is recommended by most wind-tower manufacturers. This means all components of the plant must be tied to one another in a common grounding system, grounded to earth. If your tower is a guyed pole, the guy wires themselves must be included in your grounding system. In a good wind environment, you have a highly charged atmosphere where static electricity is common. Grounding helps to bleed off the static charge, as well as to insure the reduction of electrical current when lightning strikes.

Although your wind plant itself might appear to be a grounding system, the high frequency of lightning charges tends to travel on the outside of a conductor in a more or less straight line. The concrete foundations of your tower and the anchors for guy wires are very poor conductors. Therefore, copper conductors should be clamped to each anchor point and strung in straight lines to grounding rods. Complete grounding kits cost from $350 to $375, depending on your tower's construction. You can buy the appropriate kit from your wind-plant manufacturer.

Now that we have covered all the basics of both site evaluation and suitable equipment, we need to take a look at the relationship between harnessing wind power and storing it in your 12-volt battery system.

Wind-driven generators can be very hard on batteries, as they provide a charge that varies from nothing on calm days, to terrific during storms. The potential for both overcharging and deep discharging can decrease the life expectancy of even the best batteries. We can talk in terms of normal conditions, but we must recognize that such conditions are only mathematical averages and rarely occur on any given day in the real world.

For the best estimate of your electrical generating capacity from a wind plant, you need to establish the wind's daily output in kilowatt-hours. This KW-hour

figure will vary with your generator's rated output (in watts) and the average monthly wind speed you determined in your preliminary site evaluation. Find your generator's wattage rating on the specification plate. Then take that figure and your average wind speed figure and compare them to those on the following graphs.

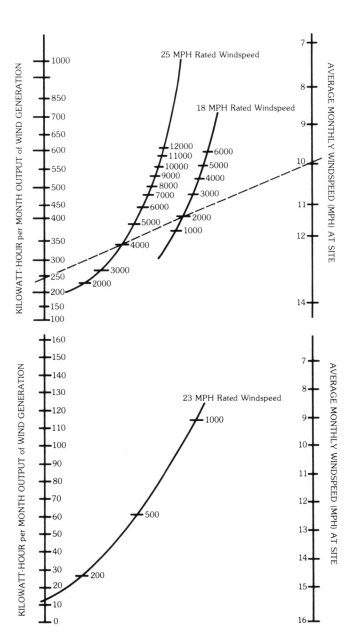

Align a straightedge from the appropriate wind speed on the right index to the rated wattage on the appropriate curve. Then read the monthly KW-hour figure on the extension of the same line where it intersects the lefthand index. Now you can divide this figure by 30 to translate the monthly average into daily output. For example, if a 1,000-watt generator (right index) operates in a 10-mile-per-hour average wind (curve), the monthly output (left index) is 100 KW-hours. Then 100 divided by 30 is 3.3 KW-hours per day.

But this figure is not yet trustworthy. You must allow for variations between the real and the ideal circumstances; include in your considerations altitude, terrain, battery condition, etc. To compensate safely for most of these variables, figure on no more than 65 percent of the rough total. In the example, 3.3 KW-hours times .65 is 2.1 KW-hours per day, realistically.

The next step is to compare your generating capacity with your storage capacity. Assuming a battery capacity of 750 amp-hours, multiply by 12 (your 12-volt system) and you find you can store 9,000 watt-hours or 9 KW-hours.

That looks pretty good, but remember again that we are dealing with artificial figures and mathematical averages. The real wind blows only intermittently, and you will need several days for the batteries to reach their maximum charge. You will probably want to increase your storage capacity to allow for about three days' worth of amp-hour usage on each charge.

Note, too, that battery storage can be too large for your charging system. If your batteries are fully discharged, all the battery cells will need time to replenish themselves. The more cells involved, the slower the charging process. Generally, a storage system should not depend entirely on a wind-generated charge, and a back-up system should be installed.

The final consideration in wind systems is a question of economics. Generally wind power is practical where average monthly wind speeds of eight miles per hour or greater prevail—on the site of the wind plant, not in the general region. When the average wind speed is about eight mph, wind-generated power costs 20 to 25 cents per KW-hour if your equipment is amortized over 15 years. Even if public utilities continue to be available at three to ten cents per KW-hour, the wind, combined with other alternatives, can save you money because now federal and state tax credits for alternative energy sources can make your amortized bottom-line cost as low as 2 cents per KW-hour. In any event, the abstract cost per KW-hour is much less significant than the cost relative to your own needs, the theory behind being your own power company.

There are smaller wind plants that generate up to 200 watts in a ten-mile-per-hour wind. They cost less but should be backed up by water, solar, or other alternative systems for greater reliability.

Naturally, the seasons of the year will play a big part in any wind plant's monthly output. During those calm months when your minimum average wind speed is too slight to keep your batteries charged, you will be happy to have a back-up system.

Water Power

Water wheels working in locations between Maine and Georgia numbered in the thousands during the nineteenth century. Used mostly in mills, their construction required both money and hard work, even without the necessity to generate electricity. Today's 12-volt technology can solve many of the problems and make an effective water system for generating electricity. But water systems still take work.

Among the important technological advances made in the last century is the water wheel developed in 1880 by Lester Pelton. The Pelton wheel uses cup-like buckets at regular intervals. A jet nozzle protruding from the end of a pipe shoots water into the buckets with enough force to turn the wheel.

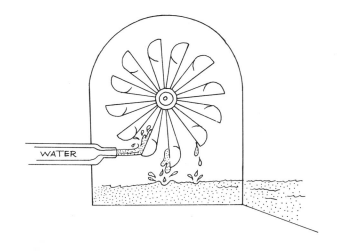

Today, commercial Pelton wheels are cast from aluminum or bronze with cups in 4 1/2-, 9- or 18-inch sizes and are usually mounted on a two-inch shaft. Manufacturers claim these wheels require a water fall of no more than 50 feet and water flow of 11.2 gallons per minute.

But before you can install a Pelton wheel or water turbine to charge your battery system, some basic principles must be understood. Just like the wind as a potential generating device, water requires consideration of some physical elements, the "fall" and "flow" just mentioned. We must measure them before knowing whether a water system is feasible on your site. And, unlike wind, water is not necessarily yours for the taking.

Water is a precious commodity, and almost everywhere in the world its use is regulated by law. Water sources generally serve many people (and other living things) at one and the same time. Restrictions govern any one person's right to take water from a communal river, stream, or lake, thereby possibly upsetting the ecological balance or diminishing the possible use of the water by others in the community. The laws generally limit water use by one of two definitions, *riparian* rights or *appropriative* rights.

Riparian water law reasons broadly that each owner of land adjacent to a body of water has a common right to "reasonable use" of the natural flow, as do each of the owner's neighbors on an equal basis. "Reasonable" means that the first use of the water on a shared basis is for domestic needs; the second use is for irrigation.

Appropriative water rights prevail at my home in California and in most of these dry, western American states. This law states that any water intended for other than domestic use must be claimed through a legal process. Just like the California Gold Rush in 1849, the first people to stake claims can take as much water as they need. Anyone who comes after them can have what is left. (Check the deed to your property; you may already have appropriative water rights.)

Outside the United States some countries have lately adopted very restrictive water laws that make the right to build private hydroelectric plants virtually impossible. Be careful if you live near state or national borders, too. The water you see may come under the jurisdiction of more than one government. In any case, check your water rights through both local and higher government authorities.

Assuming that you now have the legal right to tap your water resources for hydroelectric generation to meet your domestic needs, do you have enough water to make a plant work? Even if you live below a spectacular natural waterfall that flows year-round, you will need to build a dam or create a small pond. The pond traps the water you need at a depth sufficient for your intake pipe, with a screen to filter sand, dirt, rocks, and twigs that could damage your equipment. But do not rush to contract with a civil engineer yet. You can make some preliminary measurements of your own and probably build the dam yourself just as easily as an engineering contractor.

The operation of a water-driven generator requires water pressure, which builds under the force of gravity as a body of water moves. There are two aspects of water movement, *head* and *flow*.

Measuring the head of your water means calculating the vertical distance water falls from a dam or other source to the spot where you will install your equipment. Pick a likely spot downhill from your water source, and you can measure the head. You will work back toward the source in a primitive but very effective process.

Take a long board or straight stick, stand it upright, and mark it at your eye level. Then measure the length of the stick from your mark to the ground. Let us say it is five feet. Now stand on your likely spot with your measuring stick. Look uphill across the mark toward your water source. Pick a landmark between yourself and the source on the same level as your mark. Walk over to it. When you reach the first landmark, take another sighting and move forward again. When you have repeated the process 15 times, using the five-foot mark, you will have measured a water head of 75 feet. That distance happens to be the minimum necessary for the equipment I will describe.

If you have a friend with a stick marked at the same height, you can speed up the process by rotating on each other's landmarks, exchanging places as you move up the hill.

If you have a friend who is a pilot or has access to aircraft equipment, you can really speed up the process. I borrowed a sensitive altimeter, which I set at the 100-foot indicator. Then I simply walked downhill from the water source until the instrument registered 75 feet less. For added security, I went a few feet farther, and that is where my water turbine sits today.

Once you have found sufficient water head, you need to check the water's rate of flow, which should be a minimum of 30 gallons per minute for a turbine like mine. To measure the flow, I blocked the stream, then took a six-foot length of flexible 1 1/2-inch pipe and stuck it into the water, right at the point where I planned the dam. With the other end of my pipe in a five-gallon bucket, I checked my watch to see how long it would take to fill the bucket. It took just ten seconds. That means my rate of flow is five gallons in ten seconds, exactly 30 gallons in 60 seconds (one minute).

If either the head or the flow falls short of the minimum requirements, you can balance one against the other to a degree. Greater head permits slower flow, and faster flow can reduce the head's minimum requirement. But a head of 75 feet is still the minimum. The following chart shows the relationship.

MAXIMUM POWER OUTPUT CHART

(pipe losses not accounted for)

Head (feet)	Output in Kilowatt Hours Per Day										
	1	2	4	8	10	15	20	30	40	50	60
75								3	4	6.8	9
100						3	4	6.8	9	11.9	14.5
125						4	5.5	9	11.9	15	
150					2.9	4.8	6.8	10.5	14.5		
175					3.5	5.8	8	12.5			
200				3	4	6.8	9	14.5			
250				4	5.5	8.7	11.9	16			
300			2	5	6.8	10.5	14.5				
350			2.5	6	8	12.5					
400			3	7	9	14.5					
450			3.5	8	10.5	16					
500			4	9	11.9	16					
550		1.5	4.5	10	10	13					

FLOW IN GALLONS PER MINUTE

These minimal head and flow requirements can produce a surprising amount of practical energy if they are used to feed a remarkable new water turbine, the Water Watts. This design takes Pelton wheels one step further to absorb energy from water jets in two ways. First, the turbine's jets provide pushing action on the inside of rotating blades to turn the unit, just as Pelton's nozzle drives the cup-like buckets. But these blades are designed hydrodynamically, as aerodynamic wings are designed to lift aircraft.

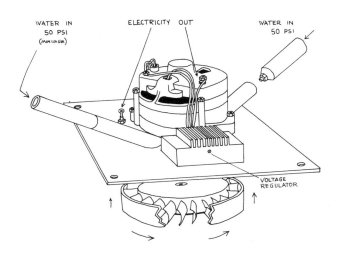

The second effect of these blades is a unique pulling action that rotates the unit more efficiently than a single jet.

When the pressure reaches 50 pounds per square inch, the turbine drives a specially adapted automotive alternator, which can produce voltage of almost any kind. The greater the head and flow, the more power this small turbine can produce. It can deliver up to 16 KW-hours a day, the equivalent of 1,333 amp-hours to store in your 12-volt batteries.

I chose the small Water Watts turbine system because of its combination of low cost with easy installation and maintenance. With 75 feet of head and 30 gallons per minute of flow, my unit delivers 3 KW-hours or 250 amp-hours a day. For current prices and updated information, please send for my catalog.

A transformer option changes Water Watts's battery-charging voltage when the turbine must be located at a distance, up to a mile, from your power storage center. Called the Long Ranger, the initial transformer changes alternator output to 440-volts AC (wild frequency). Then with another transformer-rectifier, which you should locate next to your battery system, the output is changed back to the battery-charging voltage.

As you check the suppliers of water turbines, you will notice that water flow needs are always stated as cubic feet per minute (cfm). For me, gallons per minute is much easier to understand, and I multiply the cfm figure by 7.48052 for the number of U.S. gallons per cubic foot.

Here are the Water Watts specifications. We will cover installation details in Part Two.

Alternator:
Modified Delco brand.
High-temperature stator and field windings.
Three-phase AC output for equipment options.
Tapered-roller main-bearing assembly.

Regulator:
Heavy-duty adjustable, 100% solid-state. Automatically equalizes charge at end of each heavy discharge/recharge cycle to prevent uneven charge conditions and battery-cell strain.

Turbine and Nozzles:
Turbine wheel is engineered and matched specifically for the Water Watts for maximum efficiency over a wide range of RPM. Turbine speed regulation (governing) is accomplished by selecting the appropriate size jet nozzle: 1/8", 3/16" 1/4", 5/16" or 3/8"

Canyon Industries also manufactures a water turbine, which I have not tested. They claim the ability to deliver 150 to 1,500 watt-hours a day from a flow of 20 to 30 gallons per minute. A head of 15 to 40 feet is required, and the cost is about $700. For more information, write to them at 5346 Mosquito Lake Road, Demmy, WA 98244.

I know at least one company that offers more information about Pelton wheels connected to alternators. Small Hydro-Electric Systems (SHES) asks that you send $2, and they will send you plans and illustrations for a home-constructed system. Prices range from about $500 to $2,000 for complete systems. Be sure to explain exactly what you want when writing. The address is SHES, P. O. Box 124, Custer, WA 98240.

The 12-volt hydroelectric system has been a valuable charging alternative for me during the winter months when my head and flow are at their highest. If you have water all year long, you may even be able to operate a 12-volt refrigerator or a 110-120-volt AC freezer with an inverter. If you have substantial water for even half the year, a small hydroelectric generator can be one of your most efficient charging systems. Compared to the costs of installation and monthly use for other alternatives, water is very economical. Your equipment will usually pay for itself in less than two years. After that, except for routine maintenance, your energy is free.

Increasing Efficiency

The goal of any charging system is to collect enough energy to meet our needs, and to collect it as efficiently as possible. Collecting natural kinetic energy has been limited so far to a rotary action, which interrupts the natural flow of wind or water in just one linear plane. If we could increase the kinetic effect by bending the motion (as aerodynamic design does), we might double our energy-collecting efforts and the efficiency of our systems.

In Sacramento, California, an imaginative inventor has developed an engine that does just that. Dr. Daniel Schneider has formed the Schneider Lift Translator Co. USA for increased power generation from water (hydrodynamic) or wind (aerodynamic) sources. The same principles can be applied to either energy source.

Under U.S. Patent since 1977, the Schneider Lift Translator (SLT) and Schneider Hydrodynamic Power Generator (SHPG) have two parallel axles, each with two wheels of rotating vanes connected to a unique drive and transmission system. The entire unit is placed *across* the flow of the wind or a natural river channel, rather than as a linear interruption in the midst of the flow, as with regular wind plants or water wheels and turbines. The Schneider vanes, attached to the transmission system, capture kinetic energy, causing a "lifting" motion similar to the lift an airplane wing experiences. As a result, both wheels of the unit perform the usual turning action for the generator.

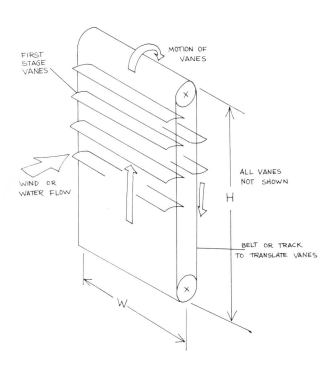

SCHNEIDER ENGINE

FIRST STAGE VANES

MOTION OF VANES

WIND OR WATER FLOW

ALL VANES NOT SHOWN

H

BELT OR TRACK TO TRANSLATE VANES

W

This principle of lift translation has attracted considerable attention from conservation and environmental groups, because the installation of a Schneider unit requires little or no damming of natural waterways. Research has shown that some Schneider equipment can produce 250 HP under a water head of no more than seven feet, with a flow of 400 cubic feet per second.

The same efficiency that protects the environment proves valuable economically as well. Maintenance is reduced because the kinetic effect is distributed across the vanes, avoiding the "centrifugal fatigue" inherent in linear energy collection. One estimate predicts that the SHPG's application to certain existing canals in northern California could reduce the cost of power by 50 to 75 percent from current utility kilowatts generated. In capturing the wind, there is considerably less danger that the SLT plant will break apart in violent storm winds.

The SLT appears to be a very good solution for providing power to small communities with appropriate geography. Recently, the director of a children's camp located on 170 acres at Point Arena, California, came to

me and asked for a recommendation on providing power to buildings in the camp. Preliminary site evaluation indicates that the location may be suitable for wind power. The SLT could be used to provide high-voltage AC power at a wild frequency. Then we can transform down this generation to charge batteries located at each building with 12-volt DC inverters sized to particular needs.

The reason for generating high-voltage AC on the wild frequency is twofold. First, with high-voltage AC, the transmission lines need not be large, which saves money on materials. Second, wild-frequency generation means eliminating the useless convention of 60-hz transmission, which requires governing a generator's speed. The governor, too, adds substantially to the cost of the installation.

Dr. Schneider has been more than generous in sharing the concept and plans he has developed. Prices begin at about $10,000 for a wind-driven 5-KW SLT. For water, the SHPG must be engineered, and prices will be quoted on a custom basis. For more information, call or write in care of Evangelyn Miller, 3691 American River Drive, Sacramento, CA 95825. Phone (916) 485-9905.

Both the hydrodynamic and aerodynamic energy sources—the wind and the water—have offered me valuable charging alternatives during the winter months. Water is especially productive for me then, when the winter runoff keeps my head and flow at their highest. But about May each year, when the rains end and the earth gets thirsty again, my system shifts over to other alternatives.

Solar Generation

Universally available to varying degrees, sunlight is a kind of energy that is permanent and free. The sun is really a nuclear plant that generates power in the form of radiant energy at an extraordinarily high kilowatt rate. Less than one-billionth of this rate, which totals a staggering 110-trillion KW-hours each day, is intercepted by the earth. Naturally, this total potential power is distributed over the entire surface of our planet.

To bring these numbers closer to home, consider that on a bright sunny day, every 11 square feet of the earth's surface facing the sun receives about one kilowatt. Even in the southwest corner of Pennsylvania, which receives one of the smallest averages of daily solar radiation (about 3.2 hours per day), a homeowner can expect potential power of about 3.2 KW-hours from the sun on an average day on a little less than 11 square feet of area.

As a matter of fact, if we could capture only ten percent of the sun shining on a square plot of land, just 100 miles on a side, we could provide the same electrical power as the present total generating capacity of the United States. One cannot help wondering how fast we might solve the current problems in solar electrical generation if we devoted the same brains, money, and other resources to solar solutions that we do to the design, engineering, and construction of nuclear power plants.

The direct conversion of solar energy occurs in two ways. "Incident" or full sunlight can be converted into heat—a process known as "photothermal" conversion—using material that absorbs the sun's rays. A greenhouse is a crude example of this process. The incident sunlight can also be converted into electricity—the simplest form being DC—through "photovoltaic" conversion. The latter process is most relevant to our needs.

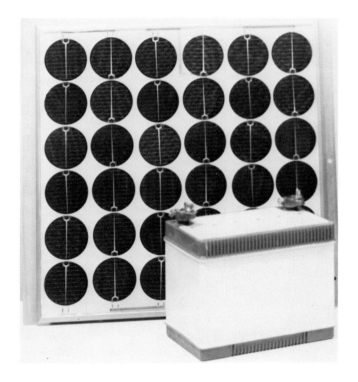

A photovoltaic device, or "solar cell," converts light directly to electricity. It is a little wafer of semiconductor material, usually silicon. The careful introduction of certain impurities into the silicon separates its positive and negative electrons from the parent atom. One side of the wafer is a surface of positive silicon, and the other side is a surface of negative silicon. There is no additional material between the two surfaces, but the invisible contact between the two sides is the key to generating electricity. When light strikes the positive side of the wafer, the negative electrons are activated, too, and they produce a tiny unit of electrical energy.

When a group of solar cells are collected to form a whole solar panel, called a solar array, quantitative electrical production is multiplied by the number of cells. A sufficient number of cells, exposed to a sufficient duration of sunlight, can create the electrical charge necessary for a battery storage system.

During the discussion of passive solar heating in Chapter 2, we noted that our collector panels must face the true angle of the sun (due south from the northern hemisphere and vice versa) for maximum heat collection. The same principle is true for panels of photovoltaic solar cells. Some manufacturers even go a step further, devising various mechanisms for tracking the sun across the sky on a daily basis and for angling them properly on a seasonal basis. Otherwise manual adjustment of the panels about twice a year is necessary for satisfactory energy collection and power storage.

New panel designs concentrate ever more cells in each unit and stress the interconnections between panels. Some units are no longer flat planes of solar wafers. They are triangular or chevron-shaped collections of silicon that depend on built-in mirrors or reflectors to soak up as much sunlight as possible within the allotted space.

Indian reservations in the American Southwest have proved to be good testing grounds for solar technology. For one thing, they get plenty of sunlight year-round. For another, they are largely subsidized by the U.S. Government, which rapidly translates into the infusion of research money through the U.S. Department of Energy.

In Arizona, the Papago village of Schuchuli became the world's first solar-powered settlement in 1978. The Papagos here depend exclusively on the sun for all their electricity. They generate 3.5 KW-hours daily under optimal conditions. That is enough power for them to operate their refrigerators, laundry and sewing machines, a 5,000 gallon-per-day water pump, and 40 fluorescent light fixtures.

The Navaho town of Sweetwater, Arizona, also uses solar energy to pump drinking water on a daily basis. They can pump 3,000 gallons at a time into a 20,000 gallon reservoir. But many Navaho homes, which are miles from the community system in Sweetwater, also use solar energy for lighting plus pumping water into their own cisterns. These individual systems maintain the charge on 80-amp-hour batteries for electricity when the sun goes down.

In the state of Nebraska, another U.S. Government installation at the Mead Field Laboratory depends on a solar array for agriculture. In the growing season, the 25-KW solar generator powers a water pump to irrigate 80 acres of cornfield. Then in the off-season, the same generator delivers its power to huge fans, which dry the crops.

These successful experiments are the product of a combined effort by the government and private industry. From the private sector, Solarex Corporation has emerged as a true pioneer in the practical engineering of photovoltaic arrays. And they found the French firm Pompes Guinard, which manufactures a DC centrifugal water pump, ideal for the combination. This kind of teamwork produced the community water-pumping system we discussed in Chapter 2.

Once again, you can call or write to Solarex Corp. for more information at 1335 Piccard Drive, Rockville, MD 20850. Telephone (301) 948-0202. Another pioneer in the field is Solar Power Corporation. You can contact them at 20 Cabot Road, Woburn, MA 01801, or telephone (617) 935-4600.

Although sunlight is free and silicon is one of the earth's most plentiful elements, the cost of manufacturing solar arrays is great. Technology continues to advance through engineering, like that of Solarex and Solar Power, but it takes a client like the U.S. Department of Energy to pick up the tab for electrifying a community.

For the rest of us, photovoltaic systems to power individual homes are very expensive. A system that can produce an average of 100 to 200 amp-hours daily costs $4,000 to $8,000. The U.S. government does offer us some help, though, by allowing a 30-percent tax credit on solar energy systems costing up to $5,000. State

governments, too—22 of them so far—offer tax incentives, but the particulars change very rapidly. I can only suggest that you check with the official agencies in your own area.

A properly planned photovoltaic generating system consists of two elements: an array of sufficient size and a battery storage system large enough to store power generated by the array. The battery storage acts as a buffer between the array and your home on sunless days and at night. Although a solar array will generate some electricity even on cloudy days, the output varies greatly on both a daily and seasonal basis. A battery system smooths out some of the variations.

The size of the array depends on your amp-hour demand. In the example we have used before, a typical home with a 12-volt powersystem, including an inverter, requires about 70 amp-hours a day. Let us say this home is located where the average daily sunlight is 4.5 hours. Now we can compare the home with the equipment available.

The most readily available and economical solar panels for a home contain 36 to 40 interconnected solar cells. Each panel produces a charge of 13.6 to 16.5 volts (1.52 to 1.90 amps, 25-31 watts). For convenience, we can round off the production to 2.0 amps per panel, then multiply by the 4.5 hours of sunlight. One panel will yield 9.0 amp-hours a day. We need about eight panels for an array sufficient to meet our 70-amp-hour demand, with one-half amp-hour left over.

However, just one sunless day will shut down our system. We must increase the number of panels to take advantage of our battery storage in cloudy weather. Actually, we will need to double or triple the number of panels.

At this point, our exclusive reliance on a solar charging system begins to look far more practical for a weekend cabin than for a permanent residence. Your use is less, and while you are in the city, the system will continue to charge the batteries. If you use the cabin Friday through Sunday, you have Monday through Thursday to restore power by solar generation.

In the example we just used—70-amp-hour demand and 4.5 hours average daily sunlight—you need only three panels for three days. But four panels will produce 252 amp-hours of charge over a seven-day period, and three days at 70 amp-hours consumes only 210 amp-hours, leaving a margin of about 42 amp-hours.

Let us back up a minute. How do we know that the cabin or the home in our example receives 4.5 hours of average daily sunlight? That figure is a recorded fact for climate zones the world over.

Here is a map of the United States that shows the average daily sunlight in all major zones. If you look at your own zone and see "4.5," you can expect 4.5 hours of sunlight on the average, per day, over the course of a year, including the winter. Of course averages are not infallible, and I have included the maps for guidance only. You know your own area better than the map. Where I live, for instance, spring always comes three to four weeks later than in the valley below me. You may find more or less sunshine than the map lists, so plan accordingly.

MEAN   DAILY   SOLAR   RADIATION   (PEAK HOURS)   ANNUAL

PUERTO RICO - VIRGIN ISLANDS

HAWAII

ALASKA

To find the average daily sunlight in other parts of the world, write to Solarex Corp. or Solar Power Corp. They both offer computerized geographical evaluations.

The total energy received from the sun in one year at any specific location is surprisingly uniform. The variation from year to year is less than ten percent. But within the year, there may be significant seasonal variations. Your average sunshine during winter months may be too little to supply your home, but a number of sunny days in a row can make up the insufficiency. Just remember that the state of charge in your battery system will fluctuate seasonally as well as daily. And think seriously about installing another charging alternative as a back-up system.

About a year ago, I installed one solar array on my roof. Last summer, after my hydroelectric system was depleted for the season, solar generation provided about 15 amp-hours a day, which was sufficient power for my low daytime electrical demands. In the evening, I plugged the power-system into my car for the additional power needed.

One great thing about my photovoltaic array is that, when I am away, my batteries are being charged automatically, and with no moving parts involved. I will be adding more solar panels as I can afford them, and with a supplementary light-duty wind plant, my system will not have to depend on the car at all in the summer.

With no moving parts to wear out, the lifetime of a solar panel should be at least 25 years. There should be no maintenance costs either. A rock or a tree limb flying through the air could damage the unit, but otherwise, I find it hard to imagine what might go wrong. So in the long run, the practical economics of solar generation get very high marks, even with the high short-term costs. And the future for photovoltaic generation looks quite bright.

Thermopile Generators

As energy sources, wind, water, and sunlight free us from our dependence on fossil fuels, but their continuous application to battery charging is filled with limitations, as we have seen. Another alternative is based on the principle of the thermopile, which produces continuous DC electricity from heat of different degrees. Small quantities of fossil fuel (natural gas, LPG, or propane) traditionally power a thermopile generator.

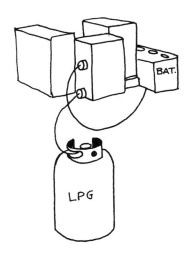

THERMOPILE GENERATOR

Thermopile is one form of thermoelectric generation, and I want to call this form by its correct name to distinguish it from other thermoelectric processes. You have probably heard about geothermal systems, which tap underground heat sources like natural steam and geysers, but that is not what thermopile is all about. I want to talk about a specially designed thermoelectric unit, which depends on heat that is produced inside a thermopile chamber.

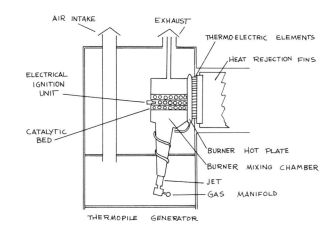

THERMOPILE GENERATOR

In this cutaway view of the chamber, we can see that gas is heated, then mixed with air in the *burner mixing chamber*. When the gas-air mixture strikes the *catalytic bed* in a burner, both heat release and low-temperature oxidation occur. The burner transfers the heat energy directly into a *thermoelectric element,* and *cooling fins* maintain a low temperature on the cold side of the element. Electric power is thus generated by maintaining this temperature difference across the element. The voltage produced initially in some models of thermopile generators is quite low and normally not practical for home requirements. So a *DC-DC converter* steps up the voltage to a more practical 12 or 24 volts.

In theory, any source of heat can be used to produce the temperatures required for the thermoelement—even a wood stove or the sun—but commercial engineering has yet to design such a unit.

Thermopile generators are completely self-contained and built to operate in any environment with no special attention or protection. This convenience makes them ideal for unattended operation over a long time period in remote areas like my wilderness home. Across the valley from me, the Forestry Service lookout tower has used a small thermopile generator for the last two years with no problems. Their system produces two amps or 48 amp-hours every 24 hours (576 watt-hours) to operate their 12-volt system for lights and communication.

Installation, maintenance, and safety of thermopile units all get very high marks. To install one of these generators, all you need is the link to your gas source on one end and two wire connections to your battery system on the other. Life expectancy for the unit is at least 20 years. The air flow can be interrupted for up to three minutes—or more, depending on the ambient air temperature—and catalytic combustion inside the chamber will resume as soon as the fuel and air mix is restored. There is no flame to burn out and no soot to be cleaned away. The only drawback I have found is the initial expense.

The Telan model 2T6 thermopile unit produced by Teledyne Energy Systems supplies 60 watts or almost 6 amps of continuous charge to a 12-volt system. That is 144 amp-hours a day, which is more than adequate for a demand in a home like the sample we discussed at the end of Chapter 2. The unit costs about $3,200 to install, and it is a one-time cost unless you decide you need increased wattage later on. Then you can parallel another thermopile unit of sufficient output to make up the difference, another one-time cost. The manufacturer claims little additional maintenance or repair.

This 60-watt unit will consume about 260 gallons a year of LPG or propane. At 60 cents a gallon, that is $156 annually or $13 a month for fuel. I already have propane in my home for cooking, refrigeration, and hot water, so there is no additional installation involved.

Now compare thermopile's initial cost and fuel costs with those for the AC generator, and remember that the thermopile unit provides constant voltage generation with no starting and stopping, and no maintenance. Thermopile begins to look even better. But when you compare it to the cost of many public utility installations and monthly bills, which are going to three figures for some folks near here this winter, thermopile generators may soon be outright bargains. Of course, if you are considering thermopile for an emergency system in the city, you will be able to use natural gas for fuel, and the ongoing cost is further reduced.

Then when you consider the 20-year life expectancy of the equipment, you can amortize the initial expense over the whole period, and your monthly outlay is only $8, plus fuel. Thermopile does not qualify for tax credits as an alternative energy source, because it consumes fossil fuels in most cases. If and when a unit is engineered to operate from other sources, such as wood or heat from the sun, that will be another story. To sum up, do not be put off by the high initial equipment costs.

Twelve-volt thermopile units are produced by Teledyne Energy Systems. For more information, you can call or write to them at 110 West Timonium Road, Timonium, MD 21093, telephone (301) 252-8220.

Just as we have seen with so many charging-system alternatives, thermopile generation's best application may be in conjunction with one or more other alternatives. The joining of gas-fueled thermopile with photovoltaic arrays is a natural case in point. With the addition of about $1,500 worth of photovoltaic panels to your thermopile system, your gas bill can be reduced up to 50 percent, and you qualify for tax credits against your solar equipment purchase. A view of the combined system is shown on the following page.

The combination includes photovoltaic panels, a thermopile chamber and fuel supply, a power-supply battery, a start-up battery, and a control box. When your sunlight is adequate, the solar panels supply an electrical charge, and the gas is shut off. Gas heating resumes whenever the sun fails to shine and the power-supply battery indicates a discharged condition. But when the

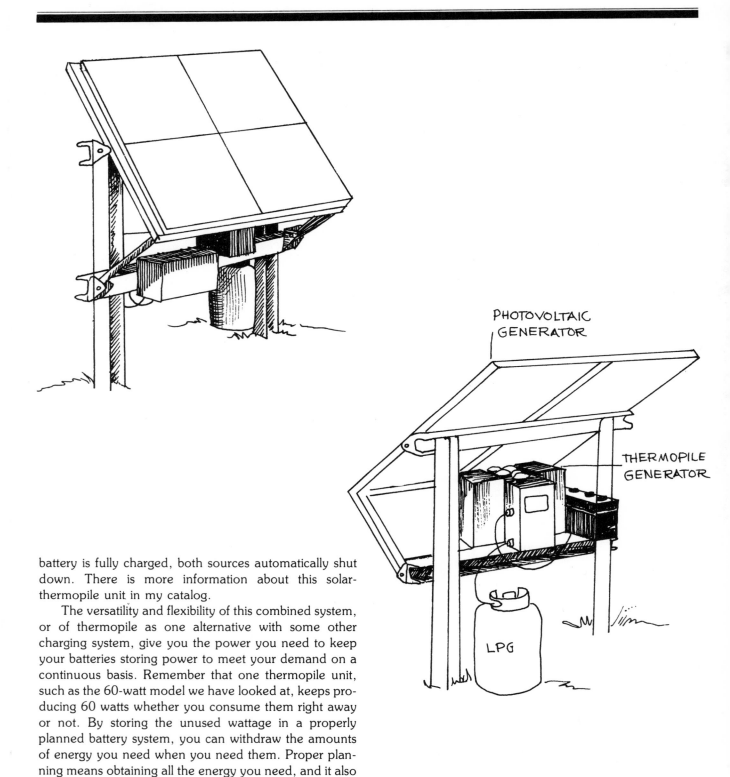

PHOTOVOLTAIC
GENERATOR

THERMOPILE
GENERATOR

LPG

battery is fully charged, both sources automatically shut down. There is more information about this solar-thermopile unit in my catalog.

The versatility and flexibility of this combined system, or of thermopile as one alternative with some other charging system, give you the power you need to keep your batteries storing power to meet your demand on a continuous basis. Remember that one thermopile unit, such as the 60-watt model we have looked at, keeps producing 60 watts whether you consume them right away or not. By storing the unused wattage in a properly planned battery system, you can withdraw the amounts of energy you need when you need them. Proper planning means obtaining all the energy you need, and it also means obtaining just enough energy for your needs.

One More Alternative

Another battery-charging system is available to anyone who owns a home or regularly visits places still served by the public utility companies. With auxiliary batteries in your car, you can mount a compact little AC-DC charging device right next to them, and they will become your portable 12-volt storage system. The other end of the charger plugs into any regular 110-120-volt AC outlet served by the utility.

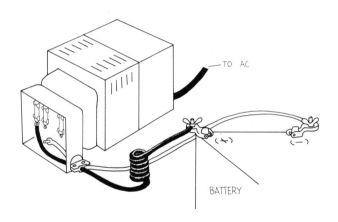

If you have a second home in the city, just plug in your charging system while you are there. When you drive back to your wilderness home, you will have stored some of the power your 12-volt powersystem requires.

If you work in a conventional shop or office on regular AC power lines, make arrangements with your accounting department first. Then you can be charging your batteries while you work during the day, storing power for the evening in your 12-volt home.

The charging unit will deliver up to 55 amps DC to your battery while drawing an average of about six amps AC from the 110-120-volt outlet. Based on a utility bill of five cents per kilowatt-hour, the weekly cost is about $6. That will keep a set of 250-amp-hour batteries fully charged to provide 70 amp-hours per day on the other end. Be sure to offer some reimbursement to the owners of the outlet you plug into.

This charger controls its own charge rate, reducing the rate as the batteries begin to fill with energy, so it will not work efficiently with the portable AC generator.

I know of two manufacturers who currently produce the compact AC-DC charging device.

- Lyle Industries
  1135 West Collins Avenue
  Orange, CA 92667
  (714) 633-2293
- Newmark, Inc.
  10648 South Painter Avenue
  Santa Fe Springs, CA 90670
  (714) 521-8210

If you cannot locate one in your RV or marine supply store, call or write the suppliers, or send for my catalog.

Time to Choose

In this long chapter, I have tried to survey all the alternatives that seem practical as charging systems for the home with an independent 12-volt powersystem and battery storage. Ultimately, you must choose the system, or systems, that are most practical for you—according to your power demands, your geography, and your pocketbook.

A combination of two or more alternatives can often contribute to the most practical and efficient total system. There is plenty of wind and water at my wilderness home during the winter and early spring, but I use my solar generator and the automotive alternator in my car for summertime support. I can do so because I use about 80 percent less power in the summer. The growing flexibility in my system is the result of new technology.

Our society has not yet brought some of this technology to the mass-produced, consumer-effective level of automobiles and generators, but maybe we can help to change that. Aside from geography and weather, the problems are simply availability and cost, problems invariably solved by the law of supply and demand.

You are designing a system with an individual output comparable to that of a giant public utility, but you do not pay for energy that your neighbors burn up while your needs are under control. If your neighbors use 500 amps and you use 20, you do not have to pay the price for the company equipment and the infrastructure necessary to serve them. This system is yours alone. It works more efficiently.

There is not much more to say about the efficient design of your own powersystem. At this point, you know the end result required, have chosen the best batteries for your needs, and know your options on charging systems for your batteries. It is time to start building.

PART TWO
Install Your System

Tools You Need

Your Wilderness Home Powersystem is installed easily and does not require many more tools than you normally have around the house. However, if you are just now moving to the country or are planning your move for the near future, you want to be prepared.

If you are not now a do-it-yourselfer, you soon will be. Please read this entire Part Two thoroughly at least once. When you have finished, there will be no surprises and your installation will go more smoothly.

As a condominium dweller in the big city, I could easily jump in the car on the spur of the moment and run to the supermarket, hardware store, or whatever. Now in my wilderness home, the run is seven miles of winding, unpaved road to the highway and another seven miles to town. I have ended too many days in a bad mood after wrestling for hours with a make-do tool on a job that would only take a few minutes if I had the right tool on hand.

Take a look at the basic collection of tools you have assembled over the years. I assume that you have a hammer and some saws. You should also have a pair of pliers and a pair of wire cutters, as well as wrenches of various sizes. Among the other tools most valuable for electrical wiring, screwdrivers play an important role. Do you have both regular-head and Phillips-head screwdrivers? You will need them. The most common size Phillips screwdrivers are #1 and #2. Screw sizes #4 and smaller will usually require Phillips size #1. Screw sizes #6 and larger use Phillips size #2. One exception to this rule is size #4 screws made in Japan, which require the use of a size #2 Phillips screwdriver.

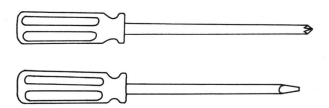

Beyond this basic tool collection there is another list of tools that you may not have needed before. The additional tools, described in the rest of this chapter, will make the installation of your power system a lot easier. All of them are available at hardware, building supply, or auto supply stores.

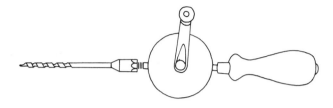

A hand drill is used for holes 1/2 inch or smaller in wood or metal.

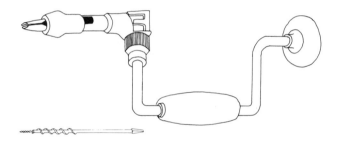

A brace and bit is for holes larger than 1/4 inch in wood only.

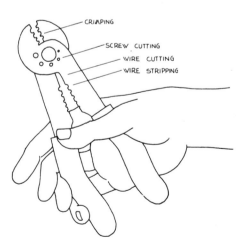

A multipurpose or crimping tool is used for cutting wire, stripping insulation, and crimping wire connections instead of soldering. This tool is also useful in car repairs and comes in two types, designed to be used on either

insulated or non-insulated terminals only. You may need both types.

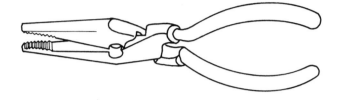

Long-nose pliers make it easy to shape wire any way you want.

A propane soldering unit is used for making the best wire connections, outside the car, that will withstand stress, weather, and corrosion, and limit your voltage drop. Practice extreme caution every time you use this unit. An open flame will readily ignite most flammable material and gases.

Soldering is not complicated. The easiest solder to use is 60/40 or 50/50 multi-core resin, which already has flux built into it. Be sure you do not use acid-core solder, because it can corrode copper wire. If you cannot find a roll of 60/40 or 50/50, you can buy the solder and flux individually. Do not become confused and buy 40/60 solder and do not buy tubes of liquid solder. To melt 40/60 solder requires 800° F. heat. Cold liquid solder sold in a tube simply will not work.

There are soldering kits available at hardware and auto supply stores from about $15 up. They contain several interchangeable soldering heads, which screw easily onto disposable propane bottles. Ask your building supply dealer to demonstrate soldering, and you will catch on after just a few minutes. If you prefer not to deal with an open flame, you might want to use a 12-volt soldering iron or one that operates on 110–120-volts AC. The latter can be used with a portable battery/inverter setup.

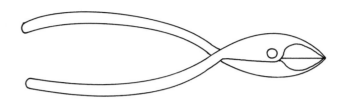

Side cutters or "dikes" are used to cut through and strip wire.

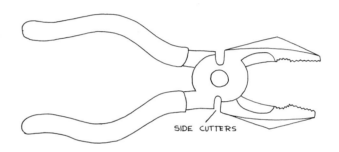

Electrician's pliers look very much like some regular pliers but are equipped with side cutters.

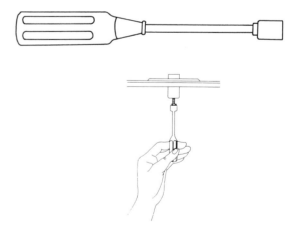

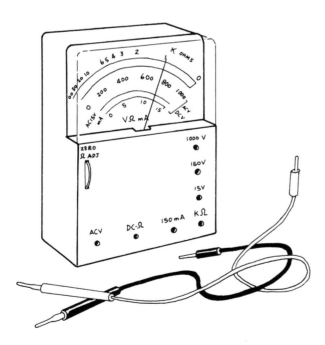

A volt/ohm meter comes with complete instructions to test your wiring for short circuits or bad connections, as well as to check the voltage output of your battery, inverter, or alternator.

A nut driver reaches nuts in cramped areas, such as the back of a 12-volt outlet installed in the floor between two close studs, etc. Some of these tools are made with solid shafts, which are absolutely useless. Be sure you get one with a hollow shaft, which will permit you to run the nut down a long screw shaft.

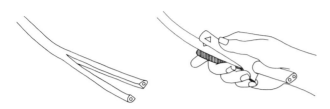

Romex strippers, if you choose Romex for your 110-120-volt AC household wiring, help to cut and strip tough plastic insulation.

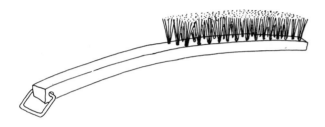

Finally, a wire brush will be most useful. You will use it to clean battery terminals and metal areas where grounding the 12-volt system to bare metal is required.

Here you have a simple enough collection of tools to enable you to install your powersystem. You will find yourself using the volt/ohm meter probably more than anything except the all-purpose wire tool. With the meter you can double-check your wiring instantly for proper connections and be sure you have a complete circuit.

CHAPTER

Wire and Wiring

Before we choose our materials, take a moment to understand wire and wiring, because, like the veins that carry your life's blood, the wiring will carry your system's energy. A properly wired installation is of the utmost importance to the long life and success of your installation.

The first thing to notice about wire is its color. Your system is 12-volt DC and will require the use of two wires to each plug, light, outlet, or switch: one black (+) for positive, one white (−) for negative or ground.

The black-for-positive and white-for-negative rule holds true in virtually all household electrical and appliance wiring. In the not-too-distant future, all color coding will be standardized. But be careful of automotive wiring. At this time it is black for negative and red for positive. Whenever you are working on your car's wiring, make a conscious mental note to double-check yourself on the color-coding. But for most of this text, we will be talking about household wiring, and the rule should be easy to remember. Some 12-volt appliances have a stripe along one solid-color wire to indicate the positive wire, and most are equipped with special plugs to make reversing the wires virtually impossible.

Why is the positive-negative distinction so important? In your 12-volt system, the direct current flows in only one direction. If you have crossed the wires in a properly fused installation, you will immediately blow the fuse. If you have not fused the installation, the wiring, outlet, or appliance could be ruined by the reversed polarity.

I mentioned the special 12-volt outlets and plugs in Part One. The most common types look similar to the cigarette lighter and receptacle in a car. When you use this type of plug for 12-volt lights and accessories you will gain some protection from reversed polarity. Your wiring behind the outlet must also preserve the positive and negative poles, as shown on the next page.

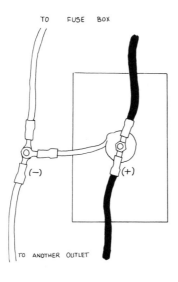

TO FUSE BOX

(−) (+)

TO ANOTHER OUTLET

12 VOLT — OUTLET

10/3 600 V

Another basic wiring distinction to keep in mind is the difference between your 12-volt DC system and the usual 110-120-volt AC system in a city household. The two operate on completely separate principles and carry different kinds of power loads. Even if your wilderness home is already equipped for 110-120-volt AC, you must add new wiring for your 12-volt system. This fact is why we are starting from scratch in these installation instructions.

If you have decided to install a power inverter, you can use the AC wiring to carry the output power, but you absolutely cannot use the same wiring for 12 volts DC and 110-120 volts AC simultaneously. Serious damage, even fire, may result.

Now let us look at the wire itself. You probably already know that copper wire is always used today because it is the best electrical conductor. Do you also know that stranded wire is the easiest to work with?

The alternatives to stranded wire are solid wire and Romex. Solid wire for a 12-volt system may not be acceptable according to the building codes, so we will not consider it. Romex is two or three solid wires encased in tough plastic as shown at the top of the next column.

Romex is used in many homes to carry AC current because no conduit is needed. If you have chosen to install a power inverter for AC appliances in your wilderness home, you may want to wire just those AC circuits with Romex. You will need the special Romex stripping tool, described in the previous chapter, to cut the heavy insulation.

For most of our wiring here, we will be talking about stranded copper wire. There are two basic types— automotive and household—and there are some important differences between the two.

Stranded automotive wire costs two or three times as much as stranded household wire. Auto wiring has a tough coating of heavy insulation to withstand the heat and stress inside your engine compartment. You can buy just the lengths you need at an auto supply store for wiring your car. But you need not spend as much for the wiring in your home.

Both household and automotive stranded wire are available in certain standard sizes. In the illustration on the next page, the different sizes are reproduced to actual scale. Note that the smaller the number identifying the wire, the larger the wire itself.

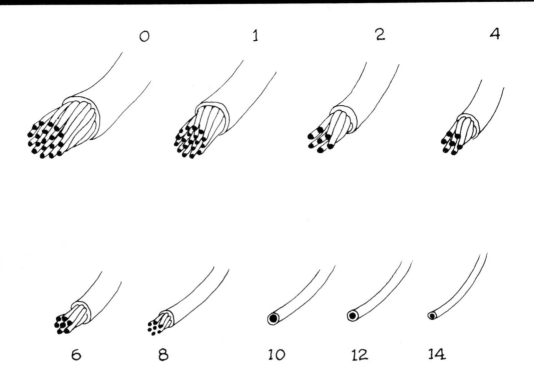

You will need to choose wire for four areas of your installation: in your car, in your house, between the car and the house, and between another charging source (such as a wind plant) and the house. All wire used in your car should be automotive stranded, #8 or larger.

The highest amperage amount you might draw at one time is important. No matter how many amp-hours of storage your batteries can provide, the main consideration here is your highest rate of discharge or withdrawal. That figure determines the size for wire running from the car through the wall to the fuse box inside your house. The higher the maximum amperage expected, the bigger the wire required. This wire should be automotive stranded to withstand weather and the stress of plugging and unplugging the house.

Wire size is also especially important if you intend to plug a large auxiliary battery in the car into a smaller in-house battery. The wire in this case must be able to handle the high amperage that will instantly surge from one battery to the other.

Inside your house, I recommend #10 or #8 wire. Someone may tell you smaller (higher number) wire will do. While smaller wire is less expensive and easier to work with, you invite annoying power fluctuations. By installing the slightly heavier gauge (#10 or #8), you minimize voltage drop in your installation.

Voltage drop is the condition you notice when you turn on another accessory and the lights dim slightly or the TV picture jumps and becomes slightly smaller. The distance you run any given size of wire from one point to the next determines how much resistance there will be to the flow of 12-volt electricity in that wire. The longer the distance, the more resistance and the greater the potential voltage drop. If you have not considered this point in advance, you may end up with 10 volts instead of 12 at the point where you plug in a 12-volt appliance.

The wire between any of the alternative charging systems and your house should be stranded, direct-burial type. It may also need to be as large as #4, depending on the distance involved. But remember, too, the transformer-rectifier system we discussed in Chapter 4. For each distance exceeding 100 feet, this system will change your voltage so smaller wire can be used.

Here is a chart that shows you the voltage you can expect to lose over a distance of 100 feet with 12 volts at 10 amps. (Another chart in Chapter 9 shows the voltage drop with 110-120 volts.) The chart uses the Greek letter Ω, which is the sign for an ohm, the unit of measurement for resistance.

#10 wire	$.100\Omega$	1.00 volt loss
# 8 wire	$.065\Omega$	.60 volt loss
# 6 wire	$.040\Omega$	.40 volt loss
# 4 wire	$.020\Omega$	.26 volt loss
# 2 wire	$.016\Omega$	.16 volt loss
# 0 wire	$.010\Omega$	.10 volt loss

When using this chart, keep in mind that, when the distance between the fuse box and an outlet is 100 feet, we are really talking about 200 feet of wiring, because two wires are needed to complete a circuit. Also keep in mind that, if your total distance is only 50 feet, you can cut the loss figure from the chart in half. Likewise, if your installation will operate at 20 amps, you will have to double the voltage loss from the chart.

The heavier the wire, the greater the cost in dollars, but the fewer your problems. Number eight wire is the best size for use inside the house, especially in the kitchen or laundry areas, where you may some day use high-amperage, 12-volt appliances. Number eight wire is also essential to reach from the fuse box to a second or third story in the house. The size of wire you choose in your home, and for use between the charging system and your home, is important for your system to function at its best.

Wire connections are the final concern in our brief rundown on wiring in general. In both the wire-to-wire connection and the wire-to-fixture (outlet, switch, or light) connection, there is the chance of contact resistance. This problem is another source of voltage drop, but more importantly, it can cause fire, because the heat generated is concentrated in one small area.

When I first began my installation, I did not know about contact resistance. I did not crimp my connections tightly enough. Later, I had to re-do my system, because the connections became loose, and I was experiencing such things as shorter fluorescent-tube life. I finally soldered most connections in the house.

On the other hand, solder can never be used to make mechanical connections in electrical wire. An example of a mechanical connection is the rivet in a Y-shaped connector. If the rivet is loose or wobbly, it should be replaced with a stronger one, then soldered to the wire. Melting solder directly onto the insufficient rivet would defeat the purpose of the moving part.

Sometimes common sense will tell you that solder is inappropriate. A properly crimped connection will be far more reliable than solder in your car, for instance. As you move over uneven surfaces, some degree of flexibility means added strength in the installation.

Crimping wire-to-wire connections with a multipurpose tool works fine, but take care to make the crimps good and tight. Terminals should be clean and free from dirt, grease, or rust before you install them.

As we begin to discuss solder and crimping with all-purpose tools, we begin to move into the next basic area of your installation. You need to make another list now, this time of the materials required to get the job done.

Materials You Need

Now that you have a complete selection of tools on hand and know your wiring, we are ready to assemble the materials for your installation. Most of them are available at hardware and building supply stores, but for much of the specialized 12-volt equipment, you must turn to mail-order, auto parts, marine hardware, RV-camper supply, and electronic stores.

Return for a few minutes to the design you produced in Part One. Either in your house or on your floor plan, you need to know the location of every light, outlet, and switch you will install. If you have not placed them all in your design, do so now. It is almost never possible to overestimate the number of 12-volt outlets you will need, so I suggest you plan as many as you can afford.

Once you have pinned down these details, start measuring. You need to know—foot by foot—how far it is from each outlet or light to its switch, and from each switch, along the wall, to the fuse box. Then you will know how much wire you need for the house.

Now you can buy twice the length you have determined, one or more spools of black positive (+) wire and one or more spools of white negative (−) wire. There is no harm in allowing some extra footage so you will not be caught short later.

These measurements should give you the total quantity of wiring you need for your in-house system. Now let us cover all the materials needed to wire your car for the system.

For wiring the car...

You start by measuring again. Inside your car, measure the distance to the auxiliary batteries, and the distance from the car to the house. When you have measured those distances, you can go ahead and buy your stranded automotive wire.

If there is no power inverter in your system, you can use #8 stranded wire within the car to the power socket and from the outside plug to the house. The #8 stranded automotive wire costs from 20 cents a foot up. With an inverter up to 400 watts, #6 wire should be used. It costs about 30 cents a foot. Inverters of 400 to 600 watts require #4 wire at about 60 cents a foot. Inverters of 600 to 1,000 watts require #2 wire at about 70 cents a foot. These larger wires will only be required from the battery to the plug-in receptacle and to the ground.

In addition to the stranded automotive wire, you will need universal battery terminals with wing nuts, one for each battery post. And you will need about another eight feet of #16 stranded automotive wire, which will be used to activate the isolator solenoid for your dual-battery system.

If you cannot locate the accessory terminal under the dashboard, or whenever you are wiring the car's cramped spaces, you may use a quick-splice or in-line tap. To activate your dual-battery isolator, use a #16-14 quick-splice. These quick taps are useful in some situations, but I do not recommend them to connect wires inside your house. When I tried them, I found they were likely to cut some of the strands in the stranded wire. In that case, voltage drop resulted.

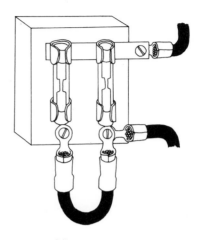

2" JUMPER

Getting back to the car, fusing the circuit will best protect the battery that powers your home. You will need one 2-position fuse block, two 50-amp fuses, a 2-inch jumper wire (black #8), and a sheet-metal screw to mount the block. The total cost is about $4.50 at an RV-camper or auto supply store.

The riveted contacts on this type of fuse block have a tendency to create high contact resistance at the rivet point. If not corrected, the heat and corrosion of resistance can substantially shorten the block's service life. In most cases, soldering all the rivet points properly will solve the problem. If your alternator output does not exceed 50 amps, circuit breakers are available and can be substituted. Cost is about $3.

Your dual-battery installation will also require a battery-isolating solenoid rated for 80 amps of continuous duty, normally open position. The solenoids usually available at an auto supply store will not work here, because they are not rated for continuous duty. The solenoid you need costs about $10 by mail or at an electronic, marine, or RV-camper store.

If your car is not already equipped with an ammeter on the dashboard, you will need one to monitor the dual-battery charge rate as you drive. An automotive ammeter will show both your charge and discharge rates in precise calibrations, unlike the model you will later install inside the house. This model costs about $10 and is available at any auto supply store.

You will also need a number of grommets, the little rubber circles you press into holes in the metal frame of the car, to prevent chafing the wires. Finally, pick up a couple of packages of wire ties and a roll of electrician's tape.

For wiring house-to-car...

Wire connections inside your car will require the same type of wire terminals used for connections inside your house, which we will cover shortly. But between your list of materials for wiring your car and your list for in-house wiring is the one big connection between the house and the car.

You need an outlet or receptacle of the proper amperage capacity on the front end of all your in-car wiring. And you want a matching plug on the end of all your wiring from the house. A specially designed Powersocket rated at 100+ amps is available from The Wilderness Home Powersystem. The company also has the only commercially available fast-charge override/dual-battery isolation system in kit form.

Then all you need is 15 to 20 feet of stranded automotive wire to reach from the fuse box in the house to your front bumper. You can wrap this wire in wire loom—available at the same store where you buy the wiring—to make a good strong cord that will withstand weather and outdoor stress.

There is one exception to this easy step. If you cannot park your car closer than 15 feet from the house, you will need to beef up the car-to-house wiring to prevent voltage drop. The solution is relatively simple. Consult Chapter 9 now, and add the materials for a junction-box setup to your list.

For other charging systems...

Each charging alternative brings its own unique wiring requirements and materials. I have taken some time here to look closely at the use of your car, because it is such a basic system and one you will probably want as a back-up for the other alternatives. The list of wire connectors and other general materials will hold true for the most part as you install any charging system. But for the specialized materials and wiring of each other alternative, turn to Chapter 9, where we can deal with them one at a time.

For your inverter...

If you have decided to install a power inverter, pause now to skim Chapter 11. Make a list of 110-120-volt AC materials you will need, then continue with the 12-volt DC materials listed here.

For wiring the house...

For the big 12-volt job inside your house, you will need all the spools of wire you measured at the beginning of the chapter. You will also need a variety of terminals and attachment devices. Your selection and the number you need are based on your own count of outlets, lights, switches, and various connections, and on the size of wire you are using. Auto supply stores are your best bet for these terminals, which usually come in packages of three or more and cost from 50 cents a package up.

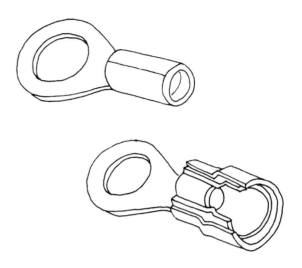

Wire terminals vary from the simple ring-and-barrel model on the left to the highest quality, insulated, sleeved, positive-crimping model on the right. The basic purpose of the terminals is to attach a length of wire to the bolt or machine screw on an electrical fixture. The wire end is crimped into the barrel of the terminal, and the ring encircles the screw like a washer.

The specific terminals you will need and their probable uses in your installation, with some idea of the quantity you need are given on the next page.

With the multi-purpose tool described earlier and these good crimp- or solder-type terminals, most of your wire-to-wire connections will be safe and secure. But there is another list of incidental wire attachments that you will be happy to have on hand.

- PVC (polyvinyl chloride) coated electricians tape
- Wire loom to wrap two or more wires together as a cord
- Nylon wire ties for securing wire to wall and floor studs, as well as in the car.

Any wiring beneath your floor, or even up in some attics, may be exposed to weather or dirt conditions that do not exist inside the walls of your home. Flexible metal tubing called electrical conduit is the extra protection your exposed wiring needs. The conduit can be the same used for regular AC wiring. Your building supply dealer can show you the various kinds available. They cost about 25 cents a foot and up.

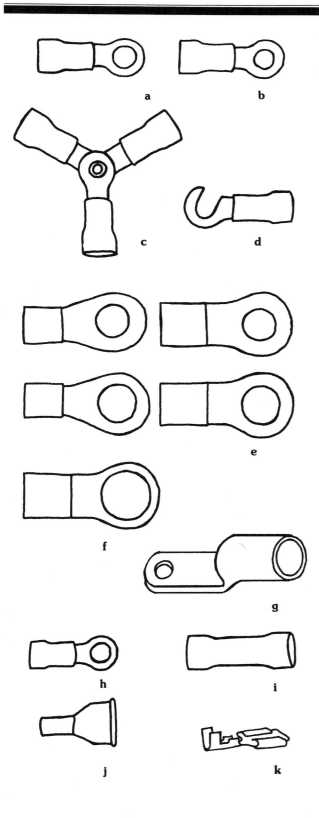

Now how many 12-volt outlets do you need? Add them up. The outlets are available at most marine or RV-camper supply stores for about $3.25 each.

Wall switches, mounting boxes, and switch plates for your lights are available at any hardware store. Twelve-volt outlets will fit into standard outlet mounting boxes, too. Each item costs less than 60 cents. Remember to pick up various sizes of wood screws for mounting your fixtures and your fuse box.

Your switches need not be labeled 12-volt, but be sure you get the snap-action type, which makes a loud click. The newer, silent type is made with mercury contacts that arc and usually fail after a short time when used in DC applications, and they therefore pose a fire hazard. The silent switch will work fine, though, on the 110-120-volt AC side of an inverter installation.

a Ring terminals: #12-10 x 8, solder type or wire-crimp type (depending on your installation), for fuse box, outlets, and in-car connections with #8 wire. At least three dozen.

b Ring terminals: #12-10 x 10, solder or crimp type, for fuse box, etc. with #10 wire. At least three dozen.

c Three-way or "Y" terminals: #10-12, wire-crimp type only. For use with 12-volt outlets. One for each outlet.

d Hook terminals: #12-10, crimp type, for connections to switches. Two for each switch.

e Ring terminals: #8 x 5/16 inch or 3/8 inch, #6 x 5/16 inch or 3/8 inch, both solder or crimp type, for use with larger bolts as in grounding to car frame, wing nuts on battery terminals, inverters, etc. The #8 or #6 refers to your wire size. Get several of each.

f Ring terminals: #4 x 5/16 inch or 3/8 inch or 1/2 inch, as above for #4 wire, especially for the junction boxes in car-to-house connection. Two for each length of #4 wire you will connect.

g Lug terminals: #4 or #2, solder type only, for special situations where heavier wire is used.

h Ring terminals: #16 x 10, crimp type, for connecting dual-battery isolator to accessory terminal. You need two.

i Butt splices: #10-12, crimp type, for connecting two #10 wires to each other. A dozen or so.

j Nylon closed-end terminals, "wire nuts," in lieu of butt splices in tight places where it is hard to use wire-connecting tools. Two for every in-house fixture.

k Male slip-on terminal: #10-12, for connecting #10 wire to new-style 12-volt outlet. (Not made for #6-8 wire, but #8 wire can be inserted by spreading terminal barrel slightly.) Two for each new outlet.

For building a battery box...

In the next chapter, you may choose to place your auxiliary battery in an on-site housing, rather than in your car. Your battery box can house an inverter as well. Consult the illustrations in Chapter 9 now, and consider the materials needed to build a battery box.

If you want a metal battery box, take your plans and dimensions to a welder or a sheet-metal shop. They can build it for you for about $100. But you may prefer to build a properly insulated wooden box yourself. The total cost is about $45, and here are the materials you need.

- 1/2- or 3/4-inch plywood
- One-inch by two-inch wooden framing
- Nails and wood screws
- Hinges and a latch or lock
- R-19 or two-inch polyurethane insulation
- Tar-like paint, made especially for battery enclosures, available from aircraft parts suppliers or battery manufacturers

A word of caution here. Some motors generate sparks and should never be placed inside your battery box. Fan motors fall into this category. Place a fan so it blows air into the box, never as an exhaust fan inside.

For severe, cold-weather climates, the battery box should be located inside your house and properly vented outside. Refer to Chapter 9 for the variations on the materials list.

For fusing your installation...

I believe that one of the most important features of your whole wiring diagram is the fuse box. We will discuss the safety and easy maintenance aspects of fusing your system in detail, but for your materials list here, you should add.

- One six-circuit fuse box
- Six SFE 20-amp fuses, plus a set of spares
- Enough 7/8-inch grommets for all the wires that will pass through the fuse box

- One ground bolt, size 1/4-28 by one inch long, fine thread
- Star lock-washer and two nuts for bolt

Be sure the fuse box you buy has insulation between the positive connections and the metal box frame, or a short circuit will occur. The cost of the box, fuses, ground terminal, and grommets will run about $15.

The fuse box and related materials almost complete the list of wiring materials you need. Now you need to list in-house fixtures. Your plans at this point should already include the number of wall- and ceiling-mounted lights you selected in Part One. You should also know the number of table lamps and other free-standing appliances that you can convert to the 12-volt system by attaching the cigarette-lighter-type plugs. Add the number of plugs you need to the materials list. They are available for less than $1.75 each at most auto supply, marine hardware, or RV-camper supply stores, or by mail.

For your power center...

You may want two final items for your complete materials list—a DC ammeter and a DC voltmeter to monitor your system once you plug it in. These devices are optional, of course, but you may feel secure in having them for routine maintenance and troubleshooting.

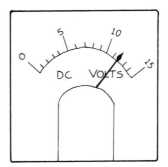
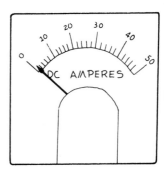

The DC ammeter, calibrated from 0 to 50 amps, will tell you how much amperage you are drawing at any one time. It differs from the automotive type in simplified calibrations for low-amp-draw reading, and it does not show a charge rate. This ammeter is available at electronic supply stores and costs from $5 to $30.

The DC voltmeter, calibrated from 0 to 15 volts, is all you need. If you decide on an ammeter, get a voltmeter that matches it for the sake of appearance. It costs from $5 to $30.

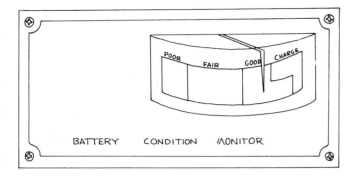

Here is another type of battery-monitor device, available at RV-camper supply stores and by mail. It is essentially a voltmeter, but instead of having calibrations, it has a green area to indicate a charged battery, yellow area for low charge, and red area for discharged. Some models can monitor two sets of batteries. They are attractive and cost between $15 and $25.

Your monitoring devices should be installed inside your home, close to your fuse box. In my house, we call that wall the "power center." I took a piece of plywood and cut it out for the faces of the meters, then stained it, and mounted it right over all the gadgets in the wall. My power center is attractive and easy for anyone at home to understand.

At this point, you have designed your system, collected the tools you need, and shopped for every piece of material required. Your installation can begin in the car or on the site of another charging system. You can start with the connection between the charging system or inside your house. We will take each as a separate phase to keep all the details in order, but remember that they are all parts of the whole system. If you have not read through the entire installation guide yet, take the time now. Whenever a question comes up while you are working, you will know where to look quickly for the answer.

Now you know where you are headed and the best way to get there, and you will be able to take great pride in the accomplishment of your finished product.

Wiring Your Car

You have several choices in your procedure now. You can begin with the wiring in your house or with any one of your charging systems. Since the car was my first system and still provides reliable power in conjunction with the other charging systems, I will describe it first.

Most electrical code requirements relating to low-voltage systems have something to say about our first and last connection here, the plug that connects your car and your house. Codes will not allow you to use a plug and a receptacle for 12 volts if the equipment is rated for 110-120 volts. The reasoning is that someone may inadvertently plug your car or your home into the higher voltage. In the beginning, I elected to go ahead and use an oddball 110-120-volt plug and receptacle configuration like this.

That plug was the first thing I could find, and I reasoned that it was very unlikely that equipment with a higher voltage output which required the same plug would ever appear in my remote area. However, this type of plug can never be used when there are auxiliary batteries at your house. The reason is that there is danger from the exposed male terminals while power is present in the system. For greater safety now, I have changed to a custom-designed car-to-home plug and receptacle, illustrated here.

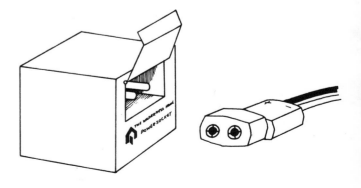

This powersocket is rated for 12-volt application and will handle demands of about 100 amps. It is code acceptable. The mounting is simple, requiring a hand drill and sheet metal screws. Then sweat-soldering makes the wire connections. The unit costs about $35, excluding wire. It is available by mail order through my catalog.

Although a 12-volt system in itself is very safe, it has a much greater shock potential in conjunction with certain parts of your car's electrical system or a power inverter. That is why it is important to disconnect your battery before attempting to install or repair any part of your system. Naturally, you never want to attempt any repair or installation when the engine is running.

Please read all these instructions thoroughly at least once and compare them with the various illustrations before starting your installation. I have illustrated with numbered instructions those parts of the installation that I believe are most complicated. Now you are ready to begin.

An open area is the best place to do the work. First, select the materials for this job from the list in the previous chapter and have them at hand. Then disconnect your batteries before starting any work.

Open the hood of your car. Locate the starting battery and remove the ground battery cable from the battery post. It will usually be a black negative cable. You will probably need a 1/2-inch wrench. Push the cable out of the way and make sure it is well clear of the battery post from which it was removed.

Most current American-made cars have a negative grounding system, but some older models, foreign cars, and large trucks will be grounded by a red cable from the positive battery post to the frame. Check to be sure which system is in your car, then disconnect the appropriate cable.

Disconnecting the ground cable first is a very important safety rule. The same rule applies to the installation of auxiliary batteries. You will not have any wires secured to both posts of a battery while you are working. Once your system is complete, the final step will be to reconnect the ground cables to the battery posts.

You already depend on the wiring in your car, and nothing you are going to do will make any dramatic changes in the present wiring circuit. We are simply going to tap into the electrical system in a couple of places to open up the unused 12-volt power and store it. You will never interrupt your car's electrical needs.

The first and simplest installation will permit you to carry power to your home from your existing car battery.

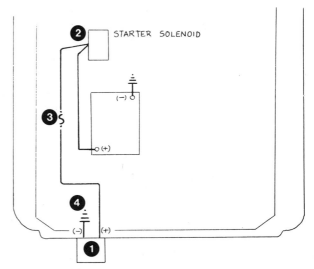

Drill two holes in the bumper or any other convenient location on the front of your car in order to mount the male receptacle (1). The necessary mounting hardware is provided with the unit.

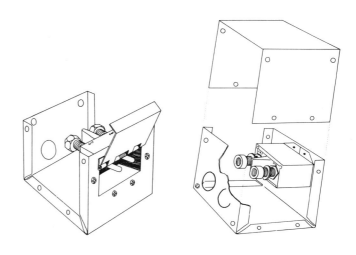

Measure the distance from your starter solenoid (2) to the receptacle on the front bumper. Cut a sufficient length of black #8 (or heavier) stranded automotive wire. Strip 1/2 inch of insulation from the end, and crimp a wire terminal onto it. Then connect it to the receptacle's positive (+) pole. Leaving the receptacle exposed, run the wire back through the grill, and connect it to the positive terminal of the starter solenoid.

Use nylon ties to secure the wire alongside nearby car wiring. Use rubber grommets in any metal holes through which you pass the wire. The grommets will prevent chafing of the wire and possible short circuits.

Fuse this wire with a fuse block (3) along the installation, wherever you can mount it most conveniently. The fuse should be rated at 100 amps, or slightly higher than the amperage generated by your alternator. Simply cut the wire, strip the ends, and crimp on the appropriate terminals. Mount the fuse block with a sheet-metal screw. Add the fuses and connect the wires.

To ground the terminal, you need to locate a body bolt inside the engine compartment (4). It should be one that makes a good tight connection to the car's frame in a spot convenient to your receptacle. Using a wire brush, clean the area down to bare metal.

Now take a short length of white #8 (or heavier) stranded automotive wire, and add terminals to each end. Connect one terminal to the receptacle's negative pole, then pass the wire through the grill. Remove the body bolt with your wrench, slide the other terminal onto it, then tighten the bolt snugly back into place against the frame.

Replace the cover on the receptacle and secure it. Your car is now wired with the single-battery power-system. You can now reconnect the battery cable you disconnected.

One caution here. You will absolutely want your receptacle on the front of the car, no matter where the battery is. If you have to back up to the house to plug into the receptacle, you will be running a serious risk of pumping carbon-monoxide fumes into the house when your engine idles. Even if you need a few more feet of wire, install your outlet on the front of the car.

That is all there is to it—as long as you can get by on a few low-amperage lights and accessories for a day or two. This installation is perfect for a small vacation cabin.

As you know by now, I firmly believe in an auxiliary-battery installation to power the home, always leaving one battery just to start the car. To gain this extra protection, we just need to tap into a few more places in the car's wiring.

Your auxiliary battery may be located either in your car or on-site at your home. You should have chosen the most likely spot as you designed your system in Part One.

The placement of auxiliary batteries in your car may challenge your creativity. Fortunately there is a variety of straps or brackets, available at most auto supply stores, that you can bolt into place wherever you choose. Another alternative is to ask a welder to make a metal box or carrier for you. Or, you can easily bend aluminum angle brackets, available from a building supply dealer, into the required shape and mount them against the car frame with sheet-metal screws.

If you have chosen two six-volt industrial batteries for your auxiliary system, you will need to connect them in series to produce 12 volts. We covered series paralleling in Chapter 3 when we discussed increased amp-hour capacity and voltage. A short length of connecting cable is used between the positive post of one battery and the negative post of the other. The length of cable is available from your battery dealer for about $7.50, and it is easy to connect, as shown here.

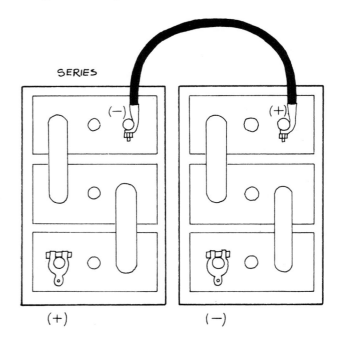

SERIES

(+)   (−)

Any battery installation in your car is subject to two important cautions. First, be sure that batteries are securely fastened down with straps or metal restraints. You do not want a battery to tip over and spill corrosive electrolyte acid when you make a sharp turn or come to a sudden stop. Second, make certain that the location you choose for your batteries is well ventilated to prevent hydrogen gas build-up, which may result in an explosion.

One good way to vent batteries is to replace their caps with chemical bottle stoppers and tubing. Measure the size of your caps, then contact a medical supply house or certain department stores that carry chemical bottle stoppers with holes in them. The same source may have the tubing, but you could use aquarium hose just as well. Run all the tubes together through a larger hole you can cut and grommet in the car's body. You can use the same technique to vent an on-site battery box.

Hydro-Catalator Company makes a special hydrocap to replace your existing battery cell cap. The hydrocap captures hydrogen gas as it is expelled during charging. Through a catalytic process, the hydrogen gas is turned back into water, which is returned to the battery cell. These caps are expensive, about $8 each, but the manufacturer claims no additional venting is needed. For more information, enclose a sample of your battery cap and send it to Hydro-Catalator Company, 3579 East 10th Court, P. O. Box 3648, Hialeah, FL 33013. The company will let you know what size hydrocap you need and the cost.

Once auxiliary batteries are installed in your car, there are four more simple installations to insure an effective system.

- Another type of solenoid to function as a battery isolator
- A fuse block to stop any overloads
- An ammeter in the car to monitor the charge
- An override switch to bypass the voltage regulator

The different type of solenoid that serves as the battery isolator closes contact when you turn on the ignition and start your car's engine. The alternator can then deliver its amperage to the auxiliary battery. But if your starting battery is low for some reason, the alternator will charge both batteries.

Now take a look at the total system shown on page 86. As you use this diagram, keep in mind that it is schematic only. The actual placement of parts indicated will be different under the hood and in your car. The designations "right" and "left" refer to the placement of elements as you face this drawing. Whenever you measure from point to point, you must use the actual installation, referring to the diagram only for the correct overall sequence.

If you have not already done so, disconnect the grounding cable from the starting battery (1).

Now you can begin with the battery isolator solenoid (2). Under the hood, find a place it will fit somewhere around the starter solenoid (3) and the auxiliary battery (4). Bolt it to the car frame.

Look at your alternator (5). On the back, find the "bat." terminal and remove the one large red wire (+). Tape the loose end and stow it away securely.

Cut a length of black positive (+) #8 stranded automotive wire. Crimp on the appropriate connectors at both ends. Then make the connections between the left side of the battery isolator solenoid and the positive terminal on the starter solenoid. If the positive terminal of the starter solenoid is impossible to reach, you may connect this end of the wire to the positive terminal on your starting battery (6) instead.

Install battery terminals with wing nuts on each post of the auxiliary battery. Then take another length of the same positive wire with connectors on each end, and make the connection between the right side of the battery isolator solenoid and the auxiliary battery.

If you intend to install an ammeter (7) to monitor the charging rate of the auxiliary battery, you will locate it on or under your dashboard and attach it according to the manufacturer's instructions. The connection will be made by interrupting the wire between the battery isolator solenoid and the positive terminal on the auxiliary battery.

Take another length of the same positive wire with connectors on both ends. Make the connection between the right side of the battery isolator solenoid and the "bat." terminal on the alternator.

Next, install the fuse block or circuit breaker (8) with a sheet-metal screw, secure it in place on the car frame close to your auxiliary battery's negative terminal. Cut a short length of white negative (−) #8 stranded automotive wire. Crimp on the appropriate connectors, and make the connection between the car frame and the left side of the fuse block or circuit breaker. You will not make the final connection from the right side of the fuse block to the auxiliary battery's negative terminal until all other installation steps have been completed.

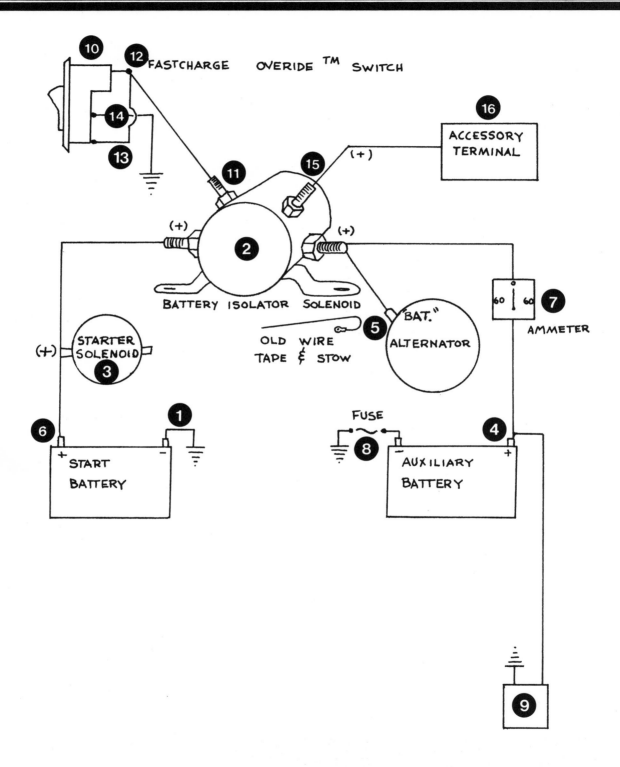

FASTCHARGE OVERIDE ™ SWITCH

ACCESSORY TERMINAL

(+)

(+)   (+)

BATTERY ISOLATOR SOLENOID

(+)

STARTER SOLENOID

OLD WIRE TAPE & STOW

"BAT." ALTERNATOR

60   60

AMMETER

FUSE

(+)

+ START BATTERY

AUXILIARY BATTERY   +

You need a two-inch jumper to connect two 50-amp fuses in parallel on the fuse block itself. Remember, if these non-mechanical rivet points on your fuse block are loose, you should solder them.

Now take a length of black positive (+) wire (size corresponding to the inverter size you may have chosen) to run from the positive terminal of the auxiliary battery to the receptacle on the front bumper (9). Crimp on the appropriate connectors, and connect one end of the wire to the positive post of the auxiliary battery. Find the most convenient path to the receptacle on the front bumper, and secure the wire along the way with nylon wire ties or grommets as needed. Then connect the other end of the wire to the positive bolt on the receptacle.

To ground your receptacle connection, select and cut a short length of white negative (−) wire of the same size as the black wire you have just installed. Crimp on the appropriate connectors. Find a body bolt nearby on the car frame, and remove it with your wrench. Connect one end of the wire to the negative post on the receptacle, and secure the other end to the body bolt. Replace the bolt, and your receptacle is properly grounded.

Now your powersocket is installed, properly fused and grounded. You have an auxiliary battery system with an ammeter to monitor the charge as you drive. The final step to insure your control is the installation of the means to maintain constant current charging, as we discussed in Chapter 4.

The key to maintaining a constant current charge is to bypass the voltage regulator on a selective basis. All you need is a switch which will give you that choice. You may be able to design one yourself, or you may wish to send for the model my friend Clyde designed especially for the Wilderness Home Powersystem, trademarked under the name Fastcharge Overide Switch. A rocker switch with a built-in light is recommended to remind you that the Fastcharge Overide is on. Mount the switch (10) on the dashboard so it can be seen and reached by the driver.

Cut a length of black positive (+) #16 wire long enough to reach from point (11) on your battery isolator solenoid to point (12) on the switch. Crimp on the appropriate connectors, and make the connection. Take another short piece of the same wire, and make the connection between point (12) and point (13) on the switch. Then cut a piece of white negative (−) #16 wire, and make the connection between point (14) on the switch

and the nearest body bolt on or under the dashboard. Finally, cut a long piece of black positive (+) #16 wire to reach from point (15) on the battery isolator solenoid to the accessory terminal (16) under your dashboard. Crimp connectors on each end of this long piece of wire, and run it through the fire wall to make the connections indicated.

If you cannot find the accessory terminal or cannot easily connect into it, locate any positive wire that runs under the dashboard to an accessory activated by the ignition. Connect your starting battery, and test this wire with your volt/ohm meter. Make sure your meter is set at the 15-volt position. Touch the positive lead from the meter to a bare spot on the wire. Touch the negative lead to the car frame, and turn on the ignition. If you have chosen the right wire, your meter will register 12 volts. Then using a quick-splice, connect your black #16 wire from the isolator solenoid to your wire. Disconnect the starting battery and continue your work.

For your final connections inside the car you will need to cut one more short length of white negative (−) wire to reach from the left side of the fuse block or circuit breaker to the negative terminal of the auxiliary battery. Be sure the wire size matches the other short piece you installed earlier from the fuse block or circuit breaker to the body bolt.

Now, at last, you can reconnect the negative cable to your starting battery and test your charging system. Start your car's engine. Your newly installed ammeter under the dash should register a charge. If it registers a discharge, reverse the black wire connections on it. If the system still does not register a charge or discharge, recheck all the installation steps.

As soon as the ammeter in your car gives you the go-ahead, you are ready for your first on-site wiring. There is only one link between your wilderness home and your car as a charging system—the plug that mates with the receptacle on your front bumper.

At this point, you must know the location at home of your car's usual parking place and whether or not your wiring must be heavy enough to protect inverter operation. Both decisions affect the size of the exterior wire connection you are about to make.

If your parking place is within 15 feet of the house, you can go right ahead with these instructions. If you must park 15 to 100 feet away, you should look now at the junction box setup described in Chapter 9. If you cannot park within 100 feet of your house, the car will be relatively impractical as a power source. In the following instructions, we will assume that you can park within 15 feet of the house.

For the sake of simplicity here, we will also assume that you have chosen no inverter or an inverter of 250 watts or smaller. In every other case, your wiring must be chosen according to the size of your inverter. Check both Chapter 2 regarding inverters and Chapter 6 on wire sizes to be sure you have the right wire size.

Now you are ready to attach the plug, which looks like this.

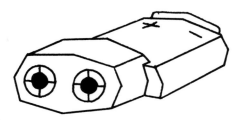

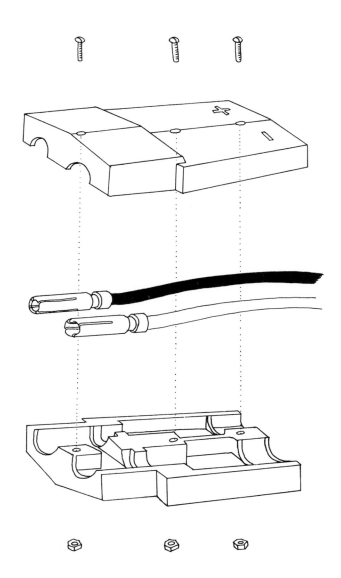

Double-check the receptacle on the bumper of your car. Be sure the positive (+) and negative (−) connections will match those on the plug.

Now cut sufficient lengths of black (+) and white (−) wire, the same size you used in the car, to reach your home. Figure on about 15 feet for each wire, 12 feet to reach from the car to the wall of the house and the rest to reach through the wall to your power center inside the house.

You will see holes in the back of your plug exactly like those on the face. These are actually hollow bolts called female jacks. To make your connection, you must disassemble the plug as shown here.

You must sweat-solder your wires into the appropriate positive (+) and negative (−) female jacks. As shown in the next illustration and described in the following procedure, you need a clamp or vise to hold the jacks in place while you heat them.

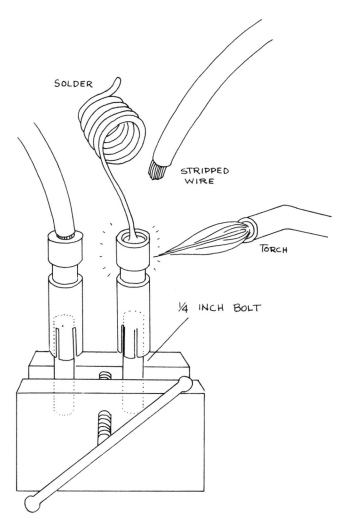

And you need either a 1/4-inch bolt or drill bit to insert into the split ends of your hollow bolts temporarily while they are clamped in the vise.

Strip 1/2 inch of insulation from the wire and set the wire aside. Using a large soldering iron or torch, heat the jack, and drip ("sweat") beads from the solder inside the jack. The wire and the jack must be made hot enough to melt the solder—450° F. for 60/40 solder, 650° F. for 50/50. The molten solder and flux will flow quickly into the joint for a cohesive bond. When the jack is filled with molten solder, remove the heat and immediately insert the stripped wire.

Be very careful of the resin flux in your solder. It will flame up if exposed to the flame or if overheated.

When both wires are sweat-soldered into their respective jacks, replace the small screws in the molded cover of your plug. Be sure the positive (+) and negative (−) marks match. Reassemble the plug and tighten the nuts.

Your plug is ready to use, but for the time being, be sure to keep it disconnected from the car. You will never plug it in until your in-house wiring is completely installed and ready to accept the reserve power from under your hood.

Now you have established the link between your car's power and your home powersystem. The plug is the key to tapping the power storage in your car and can also be used to transfer the current from the alternator to charge auxiliary batteries on-site at the house if needed.

In the next chapter, we will take a look at some details regarding the installation of alternative charging systems. These installations, just like the car system you have now installed, will lead right to the power center in your home.

CHAPTER

Plugging in Your Home

We just finished one whole chapter on the use of your car's battery and alternator system. That installation has provided a reliable power source in my home for some time now. But you already know that the car is just one of several alternatives.

In this chapter, we will look at the other alternatives, both in terms of some of the general rules that apply to all of them and some specific considerations for each system. Among the specific considerations is the question of locating the components of the system in relation to your home—some inside, some as far as a mile away. Thus we will deal with inevitable voltage drop and how to compensate for it over various distances.

By the time we finish this discussion, we will be installing our first equipment inside your home and pulling together the loose ends of all our charging alternatives, including the car, in order to bring electricity inside to your power center.

This pending union of the alternatives at your power center leads right to the first general rule underlying all our systems. *Each battery charging alternative is compatible with all the other alternatives.* In addition to an AC-generator/battery-charger combination, my water turbine and photovoltaic array are supplying amperage to the same batteries. And my car plugs into exactly the same system as the back-up power. Compatibility is one key to the overall dependability and flexibility of the system and is one goal in being your own power company.

But how is it possible? We have seen earlier that an electrical current tends to surge from a supply source to fill any uncharged gap in a system. One charging system could waste stored power by discharging electricity into another idle charging system. To insure the compatibility, we must be certain that each charging motor in the system is equipped with a blocking diode. This resistance device acts as a little traffic cop who permits current to travel in only one direction. A blocking diode is standard in equipment from many manufacturers, an option on other models. Take the time to check for this precaution when you first purchase any electrical unit for a permanent system that is both flexible and compatible.

The manufacturer's instructions that accompany any newly purchased piece of equipment provide our second general rule: *follow the manufacturer's instructions.* I will not provide here any instructions for installations different from factory recommendations. Nor will I waste

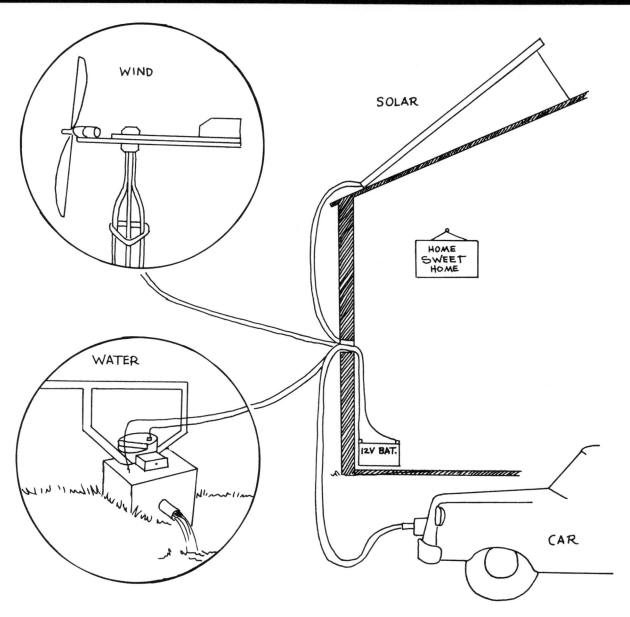

your time by rehashing information you should already have in hand. One good reason to stick to manufacturer's instructions is to preserve our protection under their warranties and guarantees.

Protection also dictates our third general rule: *follow the code.* If there are no local regulations regarding electrical wiring, follow those portions of the National Electric Code that relate to electrical systems of 50 volts or less, the second appendix in this book. I ask you to follow this code, even to exceed its requirements. Although there is

little danger of electrical shock in a 12-volt system, fire from an improper electrical installation is always possible. Naturally, a remote area would suffer badly from a fire. Correct installation is the obvious fire-preventive measure.

With these general rules in mind, we can begin looking more closely at alternatives to your car as a charging system. Suppose you do not want to wire up your car. Suppose you do not own a car. You can still store 12-volt electricity in batteries.

On-site Auxiliary Batteries

The point at which all your charging systems converge to provide the electricity you need inside the house forms the "power center" of your complete 12-volt power-system. At the end of this chapter, we will discuss your power center and its components—a fuse box, monitoring devices, and switches. The center is the ideal spot for your power storage and a good location for auxiliary batteries.

By placing your batteries right at the power center, you will save yourself many hours and a lot of wire footage when you get closer to that final connection. You will also minimize voltage drop between the power supply and the appliances and accessories you use. On-site batteries can be placed either inside your home or right outside, but there are some special considerations in either case.

If you choose an inside location, you will still want the batteries against an outside wall. During charging, your batteries create a hydrogen gas that is potentially explosive and must be vented outdoors. For appearance's sake, you will probably want to enclose the batteries in a closet or box. Gases could become trapped inside the enclosure, so venting is still a necessity.

One good way to vent the batteries is with the insertion of flexible tubing in specially adapted stoppers that replace a standard battery cap, fully described in the preceding chapter. Here is an illustration.

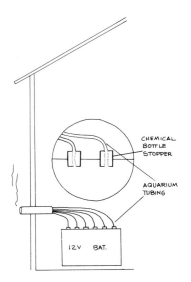

CHEMICAL BOTTLE STOPPER

AQUARIUM TUBING

12V BAT.

Another possibility is the new product from Hydro-Catalator Company described in the previous chapter.

Hydrocap*

An inside battery box will protect your batteries from extreme temperatures, to which batteries are very sensitive. But if your climate is reliably mild, you can build a little enclosure against the outside of your house and run the wiring through the wall to your power center. A layer of R19 home insulation or two-inch sheets of polyurethane all around the enclosure will help maintain the temperature inside. Insulation of an in-house battery box is also a good idea, in case you are away from home when the temperature drops significantly. Venting the battery box is a must in either an inside or an outside installation to avoid trapping these hydrogen gases inside the enclosure.

Of course you can locate your auxiliary batteries at home, even if your car is part of your charging system. The home-to-car plug we assembled at the end of the last chapter can make its first contact with your system right at the batteries, and you will have to idle the car's engine to charge your on-site batteries. While idling, there is one very important caution to observe. Remember that the terrific charging potential of your Fastcharge Overide System can produce more than 12 volts at its maximum rate. Your batteries can accept the charge, but your 12-volt fixtures cannot. To avoid burning out any part of your system, do not operate any appliances or accessories while charging. Then, when the charging cycle is complete, be certain the Fastcharge Overide is switched off.

As a back-up system, the use of the car to charge on-site batteries is okay, but it certainly is an inefficient process. You would be much better off from a cost standpoint, including the wear on your car's engine, to charge your batteries on one of the other alternatives.

Your AC Generator/Charger

Paying close attention to the printed instructions from your generator's manufacturer is extremely important, as is code wiring. You will almost certainly want the help of a qualified electrician. The potential dangers of 110-120-volt AC power are too great for guesswork.

Aside from the danger, one main drawback with a gas-powered generator is the noise it makes while operating. The relatively quiet Honda engine comes close to being an exception, but even with this model, I prefer to keep the operation as far as possible from my home.

I built a tool shed about 100 feet from the house, and that is the generator's home. Needless to say, the shed is a sturdy and weathertight structure, which also insulates

us from the noise to some degree. The generator's exhaust pipe extends through the rear wall of the shed, away from my house. I also replaced the generator's original muffler with a car muffler, which reduces the noise even further.

The real purpose of my generator, however, is not only to provide 110-120-volt AC power but to create 12-volt battery-charging power, too. Although the generator connection to my 80-amp DC charger is very simple, I have separated the two units. My charger is located at the house, close to the auxiliary batteries, for two important reasons. One is my convenience in being able to monitor the charging operation. The other is to prevent voltage drop. The 110-120-volt AC produced at the generator is stronger than the 12-volt DC at the charger. By asking the stronger current to travel the distance from the shed, rather than the weaker, my charger gets more power faster, and so do my batteries.

I laid 100 feet of #4 direct-burial stranded wire in an 18-inch deep trench between the shed and my house. That is a large wire size, but it offers maximum voltage and amperage with minimum voltage drop, regardless of the AC or DC demands I make on it, now or in the future.

The whole charging system looks like this.

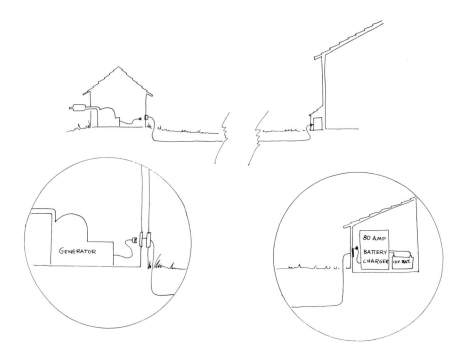

In the tool shed, the end of the buried wire comes up to the generator. I installed a regular AC plug on the end of the wire, matching a wall-mounted AC outlet from the generator itself. On the other end of the wire, at the house, I installed an AC outlet, which receives the AC plug on the battery charger.

The DC side of the charger came equipped with two wires, complete with clamps. The red positive (+) wire clamps on the battery's positive terminal, and the black negative (–) wire clamps on the battery's negative terminal.

That is all there is to it. Once you have made the four wire connections, start your generator and turn your battery charger to "high boost." The charger should deliver the maximum amperage your batteries will accept.

But suppose it does not. Some of the possible reasons why follow. You would be well advised to think about these conditions even before beginning your installation.

The ammeter on your charger will tell you how many amps are being delivered to your batteries. If an 80-amp charger shows a charging rate of only 20 or 30 amps, it could be that your batteries still have a charge left on them. The charger will not produce greater amperage than the batteries will accept.

On the other hand, the ammeter may show zero, meaning it is delivering no charge at all. The all-too-common reason may be that your batteries are fully discharged. You should never allow this condition to occur, because it will really shorten your batteries' lifetimes, but you may be able to solve the condition the first few times it happens. If you set the charger's rate on "high boost" with fully discharged batteries, the charger will shut itself down. Reset the charger on "low" for a couple of hours, and it should operate.

At the same time, take a hydrometer and check the electrolyte in these low batteries. As the level begins to come back up, you can reset the charger to its medium rate for a couple of hours, then again to "high boost" until the electrolyte reaches a reading indicated by a green area on the hydrometer.

Adjusting the charge rate in conjunction with reading your batteries' condition will solve most routine problems. But if your charger still fails to deliver, the problem may be with your installation itself.

Use your volt/ohm meter to check for voltage drop. Set it on the 110-120-volt AC mode. Check the voltage at the outlet on the generator. Then check the other end, at the outlet into which the charger is plugged. The voltage at the two points should be the same or only slightly less on the charger end. If there is significant drop, either the distance is too great or the wire is too small, or both.

The installation of a large wire size over a long distance involves the same junction box setup we will discuss in conjunction with the parking place for your car—but for a different reason. With the traditional generator, we want to preserve 110-120-volt AC power, not 12-volt DC. To use standard AC plugs and outlets on either end, we need to reduce the wire to a standard size that will fit the plugs, and the junction boxes permit the transition.

There is one further caution in a gas-powered generator system. Remember that the engine requires regular maintenance, just like a car. Unleaded fuel will burn cleaner, reducing maintenance and extending engine life, but you need oil, too—and frequent oil changes. Most portable generators do not have oil filters. You should change the oil after every 25 hours of operation—regardless of the oil maker's claims. The more expensive oils will build up more expensive sludge, but the generator still needs flushing and fresh oil every 25 hours. A proper maintenance schedule is good preventive medicine.

With proper attention to all these details, the old-fashioned AC generator can make a valuable contribution with its new battery-charging role.

Plant for the Wind

If you followed the site evaluation guidelines and design considerations in Chapter 4, you have chosen the most suitable equipment for your own aero-electric power-system. Now as you read the instructions that accompany your equipment, you can see that you certainly will need a number of people to help with this project. Otherwise, the actual connection from the wind generator on top of your sturdy tower down to your battery system is quite easy. You just make two connections on each end.

Your wire size will be determined by the distance from the wind generator's electrical connection at the top of your tower to the battery location. The length of the wire is probably the most important consideration here, including both the horizontal and the vertical distance it must travel. Your wire will run down the center of the tower and be secured to the tower by clamps. Then it will leave the tower at one of two points. You can suspend the wire from poles high enough over the ground so that it will not be an obstruction for people and cars. Or you can bury the wire in an 18-inch-deep trench running to the house. I prefer the latter for environmental reasons.

The total length of your wiring, including the height of your tower, may easily exceed 200 feet. When that is the case, you invite voltage drop, even with wire size as large as #4. The solution in this case is a DC-DC converter and switching regulator to boost the voltage and amperage delivered to your batteries. This same solution serves photovoltaic systems, and we will discuss it in that section of this chapter.

Once you have considered wire length and protection against voltage drop, you and your friends can begin to raise your tower. As this project proceeds, section by section, be sure the resulting structure will be securely

stabilized. Then you can climb back up to connect your wires.

Connect one black positive (+) wire and one white negative (−) wire to the appropriate terminals on the wind generator atop the tower. Then climb down slowly, clamping the wires to the tower at frequent intervals. Back on the ground, connect the other ends of the same wires to the appropriate terminals on your battery system. Your wind-powered electrical system is in place.

Wattage from Water

Once you have found sufficient water head and flow on your property, according to the process described in Chapter 4, the installation is very simple. There is some physical labor involved as you build your dam, then lay and cement your pipe. Otherwise, the instructions form another step-by-step procedure. First, take a look at this overview of a complete Water Watts installation to orient yourself.

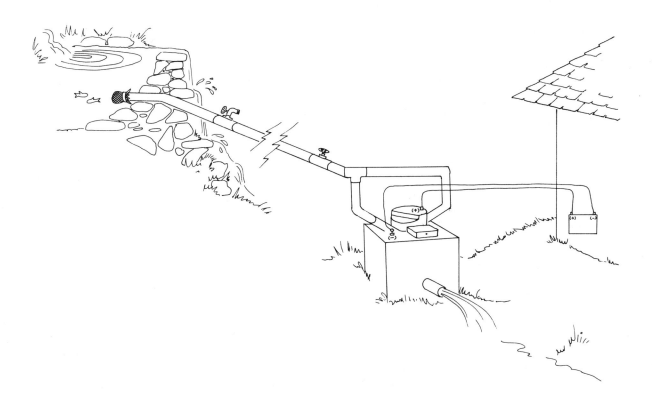

The first phase of your installation is at the water source, building a dam equipped with a screened main feed pipe. Second, you need to jump ahead to the housing for the Water Watts turbine—a box and discharge system. Then you can return to the relay pipe between the main feed and the turbine, ending with the installation of the Water Watts unit itself.

The connection between the Water Watts alternator and your power center, not unlike other systems, may require changing voltage to compensate for voltage drop, which I will explain in a few pages.

To begin the first phase, you may want to call on several friends with shovels or one friend with a big earth mover. You need to dig out a bowl at the spot on the stream you have chosen for a dam. To protect your system from rocks and debris in the stream, this bowl must be deeper than the point at which your main feed pipe will enter the flow. The pipe will be inserted about halfway up the front of the dam, permitting rocks and debris to settle in the bowl below it.

As a further precaution, build several rock baffles upstream to trap sand as well. If you have a stream bed that is dry in the summer, the baffles are especially important, because a great deal of sand and rock will be swept downstream by the first heavy rains.

When your bowl has been dug out, start gathering nearby rocks for your dam. Stack them on the downstream side of your bowl, using ready-mix cement to mortar the spaces between the rocks. When your mortared rock pile is halfway to the necessary height, lay a 20-foot length of pipe into the wall.

This first length of pipe will become the source of all the water power in your system. To protect it .from debris, cap the underwater end of this pipe with a layer of 1/4-inch mesh screen. Then while you are still working on the dam, enclose the inlet of this pipe in plastic kitchen wrap secured with rubber bands.

When your dam is completed, turn to the site you have chosen for your Water Watts turbine. The location should be as close as possible to your house and battery. You want to be sure you have as much water pressure as possible for the turbine operation. The minimum requirement is 50 pounds per square inch at the nozzle jets that turn the turbine.

The turbine will be mounted inside a caulked wooden box, through which water discharged by the system will flow. You need to build the box before laying the pipe that leads to it. Simple manufacturer's instructions for the box call for short pieces of 2-inch by 12-inch lumber nailed together to form a cube, open at the top. The metal plate of the turbine will enclose the top of the box.

Water will pour into the box through the turbine's two nozzle jets, so you need a quantity of waterproof caulking inside and a two-inch round hole on the downhill side of the box. When you insert a short length of two-inch pipe in the hole, it becomes the discharge pipe for water passing through the turbine.

You are ready for the next phase of the complete installation, laying your pipe. The main part of this procedure is covered by the instructions that come with your Water Watts unit, but I have a few suggestions.

The kind of pipe is important. I used polyvinylchloride (PVC) pipe even though a cheaper, more flexible pipe is available. I wanted a strong and durable system. PVC pipe is made in 20-foot lengths, which can be joined together with PVC cement. Each time I joined two lengths of pipe, I cleaned the surfaces at the joint thoroughly with PVC solvent to assure a good, snug joint.

The size of PVC pipe is critical in order to maintain the water pressure necessary to drive the turbine: 50 pounds per square inch. Laying pipe over a distance up to 300 feet calls for 1 1/2-inch diameter, schedule-160 pipe. You can use the same diameter for lengths to 400 feet, but increase the quality to schedule-200. For a distance of 400 to 1,000 feet, increase the pipe diameter to two inches. Figure on paying about 60 cents a foot for PVC pipe.

In addition to the pipe itself and matching gate valves to start and stop the water's flow, you need pipe tees with plugs or faucets at regular intervals. These devices permit you to vent the main feed pipe and dissipate air bubbles that form when you first admit the water. They should be installed about every 100 feet and again just before the water reaches the first gate valve at the turbine-alternator unit.

While the PVC cement dries on the pipe you have laid, you can turn your attention to the other side of the installation. Here again, the electrical connections are very simple—black positive (+) wire connecting the positive terminal on the Water Watts alternator to the positive terminal on your battery and white negative (−) wire between the corresponding negative terminals, but these wires should not be finally connected until all other steps are complete. Here, too, you may be liable to voltage drop if the total distance the wire must travel exceeds 100 feet. The transformer-rectifier installation described shortly in this chapter will solve the problem.

The final phase of your hydroelectric installation is the connection between your main feed pipe and the turbine-alternator unit sitting on your discharge box. Then remove the kitchen wrap over the inlet pipe. The Water Watts manufacturer, like all good suppliers, provides full instructions at this point. If you follow them carefully and remember the suggestions I have offered here, your water-driven charging system will be ready to go.

Solar Voltage

Before installing any photovoltaic equipment, you may want to look at some of the most relevant applied research I have seen. The technical staff at Solarex Corporation has published a very comprehensive book called the *Guide to Solar Electricity.* It is a paperback, filled with photos, charts and installation advice, selling for $6. You can order it by writing to Solarex Marketing Manager Bob Edgerton, 1335 Piccard Drive, Rockville, MD 20850.

The trickiest part of your solar installation is positioning the arrays for exposure to the maximum sunlight. You will recall some details about exposure from our earlier solar discussions in Chapters 2 and 4, but now you need to find the proper angle to tilt your panels on a seasonal basis. One effective way to find this angle is with a short tube or length of pipe, about two inches in diameter and three feet long. At noon time or close to it, in the bright sun, take your pipe outdoors to the spot where you will install your array as illustrated at the top of the next column.

Aim one end of the pipe directly at the spot for your installation, and point the other end at the sun. When the sunlight shining through the pipe makes a sharply defined circle on your installation spot, you have found the correct angle. Secure the adjustable brackets on your array, according to the manufacturer's instructions, at exactly that angle.

This position for your array will serve you well for three to six months. Then you should take your pipe out-

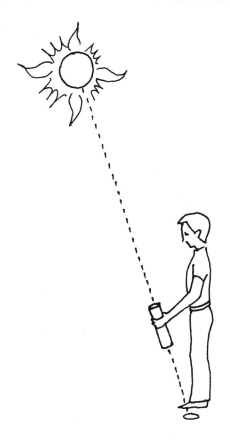

doors again at noon on a sunny day. Using the same procedure, adjust the angle for the seasonal change. Unless you have indulged yourself in the luxury of automatic tracking, you will want to change the angle at least twice every year.

The rest of your installation is very simple, if you continue to follow the instructions provided with your array. Make the two standard wire connections—black positive (+) wire between positive terminals on your array and your battery system, white negative (−) wire between corresponding negative terminals. Your wire size will be determined by the distance between your array and the battery system.

More often than not, your solar array will be located on the south side of your home's roof or on the ground close by, but as more people realize the potential of photovoltaic generation, technology will improve and greater distances may be involved. Long-distance installations are not really efficient for small arrays of just one or two panels, but for larger arrays, some means of transmitting the voltage back from an ideally sunny spot can be justified.

You can locate an array of five or more panels at a distance of 50 to 300 feet away from the battery system with proper sized wire, special wiring connections, and a device known as a switching regulator. Normally, solar panels are wired in parallel to maintain voltage and to increase amperage. But in this case, the panels should be wired in series to increase the voltage. We saw the same principle applied to batteries in Chapter 3.

With series wiring, five panels, each producing 2 amps at 16 volts, will produce 80 volts. The 80 volts can travel farther, with less voltage drop, than 16 volts in the same wire size. At the batteries, a switching regulator will return the voltage to that which is required for normal battery charging and boost the amperage.

For longer distances, up to a mile, the installation of large solar arrays involves many of the challenges already considered in typical installations of wind plants. In both cases, a DC-DC converter is added to the system. The converter boosts the output to several hundred volts or more, depending on the total distance current must travel. The switching relay at the batteries still reduces the current to normal battery-charging voltage.

All these solutions to transmitting power over extraordinary distances are custom installations. If you need more information about any such special applications, please write to me and include your specific requirements and any exact figures involved. I will be happy to offer individual recommendations and approximate costs.

If your needs are less extreme, the standard photovoltaic installation will provide one of your simplest and most effective charging systems.

Charging from a Thermopile

The thermopile battery charger you buy is a self-contained unit with no additional parts to assemble. A simple pipe connection to the fuel source on one end and two wire connections on the other end will complete your installation.

More than likely, the fuel source for your thermopile burner is already on your site for cooking and refrigeration. Simply tap into the butane or LPG line that already serves your kitchen, and run sufficient pipe to the burner in a code-approved manner. The placement of your unit is likewise simple; it should be somewhere between the fuel tank and your battery storage area.

Most building codes specify that fuel tanks be kept at least ten feet from structures or any other mechanical apparatus. Although the codes have yet to deal with thermopile units specifically, they do deal extensively with gas-installation procedures. The ten-foot rule seems like a reasonable guideline. Remember that combustion of a sort does occur inside the burner chamber. For the same reason, a thermopile should not be enclosed with your batteries, whose hydrogen-gas build-up demands caution. But the closer you can place the thermopile unit to the battery enclosure, the more efficient your charging installation will be.

Once the burner is in place and connected to your fuel tank, connect the black positive (+) wire, sized to correspond with your maximum amperage demand, between the positive terminals of the thermopile generator and your battery, then the white negative (−) wire between the corresponding negative terminals. Since your simple installation is complete, you can even go ahead and secure the grounding terminal and fire up your thermopile-charging system.

As we saw in Chapter 4, this simple thermopile installation is readily compatible with a solar array or any other charging alternative. A sensing battery with which the burner is equipped will signal the burner to begin operation only when no other charging system operates and the stored amp-hours in the power supply batteries drop below a minimum level.

One last point, which also underlines the simplicity of thermopile, is a restatement of the principle of generator location. There should never be a need for junction boxes or transformer-rectifier installations to change the thermopile's charging voltage. But for all those cases where a change in voltage is necessary, let us take a look at some of the means available.

Between Charging and Storage

Any wiring installed between your charging system and the power center in your home will be subject to some stress, whether it is exposed to the weather or buried underground. For underground power paths, you can and should use the heavy-duty direct-burial wire we have mentioned several times. For exposed wiring, protection is almost as simple.

Remember the plug from your car we left hanging on a post at your usual parking place? The wire on which that plug is mounted will carry the full power load from the car batteries to your home. It should be the heavy-duty stranded automotive wire of a size corresponding to your inverter demand. Now you can wrap your wires from the house to plug with wire loom to create a strong cord that will be impervious to outdoor stress. You may want to buy a hook to hold your plug and wire assembly off the ground, or fashion your own hook from a coat hanger, a scrap of wood, or whatever. Wrapping any outdoor wire assemblies with wire loom is essential, as is the use of wire with heavy-duty insulation on all outdoor lines.

Now let us take a look at the distances involved. We have seen that a thermopile generator need never be too far away from the batteries it charges, and we have discussed an innovative new technique to maintain solar voltage over a necessary distance. What about the other alternatives?

If you must park your car 20 to 100 feet from the power center, you need a junction box setup to minimize voltage drop. If the distance is more than 100 feet, junction boxes will not help. Solutions like the DC-DC converter we discussed or the transformer-rectifier plan that follows will not help either. They are so involved and expensive that they defeat the simplicity of battery-charging by car. So establish a definite parking place within 100 feet of your house; then the junction box setup will serve you well.

The setup consists of two standard junction boxes, one at the power source, the other at your power center. The boxes are weathertight with sufficient lengths of positive and negative #4 direct-burial wire between them. The installation looks like this.

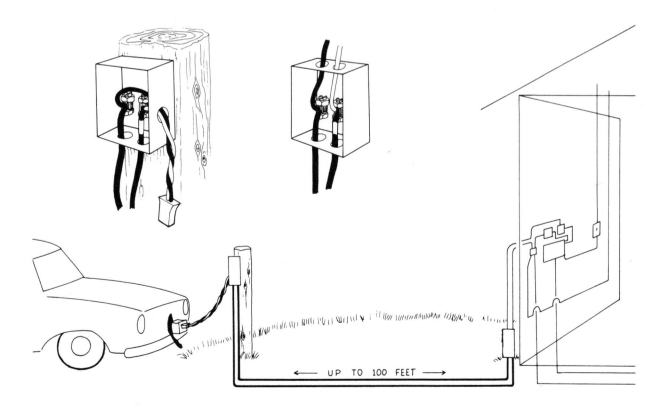

UP TO 100 FEET

You can buy two regular waterproof junction boxes at hardware or building supply stores, priced from about $7.50. You will also need four copper bolt connectors (two for each box), which cost about $1.50 each. These bolts have holes in the center for connecting large wires. Then you must get enough #4 direct-burial wire to run between the two boxes, 18 inches deep.

The #4 wire is available in black only, so buy double the length for your positive and negative wiring. Identify the negative wire with a piece of white tape at both ends.

Install one junction box on the outside wall directly opposite your power center in the house. The other will be installed on a post or tree at your usual parking place. You can let the plug hang from the junction box or the hook you have devised, but keep in mind that you will need about six feet of #8 wire (or larger according to your inverter) to reach the car.

When your junction boxes are securely mounted with wood screws, turn to the #4 wire. Leaving enough slack at either end, stretch the wire between the two boxes, and place it in a trench at least 18 inches deep. At the junction box near your parking place, match the positive wire from the car's plug to the end of the positive burial wire. Then match the ends of the corresponding negative wires. Insert each pair in the appropriate positive (+) and negative (−) bolts in the junction box, and tighten up the connection.

On the house end, take the two short lengths of #8 (or larger) wire for the inside connection, and match them to the two ends of #4 wire from the trench. Insert these pairs in the appropriate positive (+) and negative (−) bolts, and tighten them up. Run the household wires through the wall to the spot inside where your fuse box will be mounted.

Check your whole junction setup with the volt/ohm meter to be sure you have a continuous circuit from the car's plug to your power center, then insulate the copper bolts in both junction boxes with electrician's tape. Cover the boxes and weatherproof them. Then fill in the trench to bury your #4 wire.

For your AC generator out in the tool shed, the same procedure will provide junctions for outlets and different wire sizes, as we mentioned earlier. But using buried wire and junction boxes with 110-120-volt AC from your generator to your battery charger does not involve so severe a distance limitation. You could probably stretch it up to 300 feet with only minimal voltage drop using #4 wire.

Another key difference between junction boxes for the car's DC system and the generator's AC power is the number of wires you will lay in the trench. The car requires only two wires: one positive and one negative. The generator takes three: one positive, one negative, and one additional wire between the grounding terminals on your AC plugs and outlets.

Junction boxes and proper wire size can maintain adequate voltage over the short distances from your car or AC generator to your home. But a water-powered system will very often involve far greater distances.

When your water turbine must be more than 100 feet, up to one mile, from the battery storage at your power center, you need a transformer-rectifier system to preserve current transmission. The Long Ranger Option for the Water Watts turbine is one such system. It can be purchased independently through our mail order catalog.

Take a look now at this chart, which shows wire sizes for different voltages, depending on the distance from your battery.

WIRE SIZE CHART

Distance from plant to battery (feet)		0	1	4	6	8	10	12	14
10	12V						50	30	16
	110V								
20	12V				60	40	25	13	8
	110V								
50	12V		80	40	25	16	10	5	3
	110V							15	9
100	12V	50	40	20	13	8	5	3	
	110V						15	9	6
150	12V	30	27	13	8	5	3	2	
	110V					15	9	6	4

WIRE SIZE CHART

Distance from plant to battery (feet)		Wire Size (AWG)							
		0	1	4	6	8	10	12	14
200	12V	25	20	10	6	4	3		
	110V					12	9	5	3
250	12V	20	16	8	5	3	2		
	110V				15	9	6	4	2.5
300	12V	16	13	7	4	3	2		
	110V				12	9	6	3	2
400	12V	13	10	5	3	2			
	110V			15	9	6	4	2	1
500	12V	10	8	4	3				
	110V			12	9	5	3	2	1
600	12V	8	7	4	2				
	110V				12	6	4	2.5	1
800	12V	7	5	3					
	110V			15	9	5	3	2	1
1000	12V	5	4	2					
	110V	15	12	6	4.5	2.5	1.5	1	
1200	12V	4	3						
	110V	13	10	5	3	2	1		
1400	12V								
	110V	10	8.5	4	3	2	1		

MAXIMUM CURRENT (AMPS) AVAILABLE FROM HYDROELECTRIC PLANT.

My water turbine sits about 800 feet from the house. With the Long Ranger's transformer connected to the turbine's alternator output, I can transmit 440 volts AC back to my home's power center. The transmission of this high AC voltage is three-phase, so I have installed 800 feet of #14/3-600 volt Romex to deliver the power.

When it reaches the power center, a transformer-rectifier changes the high voltage AC back to battery-charging, DC voltage.

The 440-volt AC transmission is on a wild frequency, not stabilized to 60 hz. Some people have asked why I don't simply change it to 110-120 volts AC and put a governor on the frequency. My answer is that I have neither the money nor the engineering know-how to do so. The Long Ranger components cost around $350, and you can write to me for details.

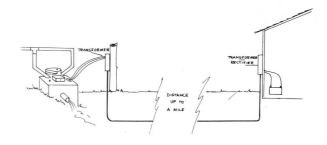

Now you know how to install at least half a dozen types of charging systems, and understand several ways to carry their power to your home. You are ready to begin plugging in your home.

Your Power Center

As the last step in wiring your charging systems, the power center is also your first in-house installation. Your fuse box is the first place where all the positive and negative wires from your battery or batteries actually connect with all the wires for the circuits or power paths in your home. I will discuss the fuse box and your monitoring system in this chapter, because they are so basic to the rest of the in-house system you will learn how to install in the next chapter.

If you can locate a junction box, transformer-rectifier, or DC-DC converter with switching regulators, along with auxiliary batteries, your power inverter, water pump, monitoring meters, and your fuse box all within a few feet of each other at a true power center, you will gain maximum comfort and control in your power-system. For even greater convenience, you may want to add a switching device.

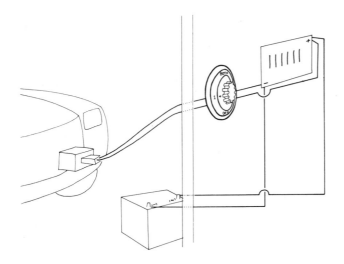
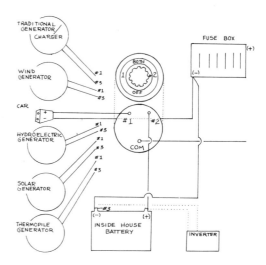

You can buy your switch at most RV-camper or marine supply stores for under $50. Although good installation instructions come with the switch, you may find the diagram at the top of the next column helpful.

There are four positions on this switch. On *1,* your car or battery #1 alone powers your home. On *2,* the auxiliary battery or battery #2 alone powers your home. With a car plugged in and the engine running (or an independent charging-source operating), turn to *Both* and you will have power to recharge all your batteries. Turn your switch to *Off* to disconnect all functions when you leave the house.

Exercise caution when using this switch. Remember never to operate 12-volt appliances or components while using a car on Fastcharge Overide to charge your system.

With both an inverter and this switching device on your system, the inverter should be connected directly to the battery, as indicated in the illustration. This placement prevents the inverter's initial power surge from severely affecting your 12-volt lighting or other sensitive accessories.

Now we are ready to install the fuse box at the spot you have chosen for the power center in your wilderness home. I cannot overemphasize the importance of fusing your whole installation. The fuse box will be your best guarantee of both safety and convenience.

There will be six power paths emanating from the box, each carrying a flow of current or 12-volt electricity. Each path is protected by a fuse, located in the circuit box, which will blow ("self-destruct") if there is a short circuit or overload—too many accessories operating at once or crossed wires—in the path. A blown fuse also protects the system from an accessory that is not working properly. All circuits in your system must be fused.

If you have a very large home, or you plan to operate a great number of appliances and accessories, you may require fusing for more than six paths or circuits. You can easily cover your installation with a second fuse box, wiring up just as many of the additional circuits as you need.

By the same token, if you have a smaller home, or very few fixtures, you may use only three, four, or five of the fuses in the basic six-fuse box. The actual positive-to-fuse and negative-to-ground wiring is the same, regardless of the number of circuits.

An example of the wiring in your fuse box is shown at the top of the next page.

The black positive (+) and white negative (−) wires that run parallel to each other throughout your whole installation will be separated in your fuse box. All negative (−) wires connect to the ground terminal block (1).

If your fuse box has no ground terminal, you can use a #10 x 1-inch bolt with a star lock-washer and two nuts to fit. Drill a hole in one corner on the bottom of the fuse box and insert the bolt, which will become your ground terminal block. Solder ring connectors to all your white wires, including the one from your power source, and slide them on the bolt. Then tighten everything into place with the washer and nuts.

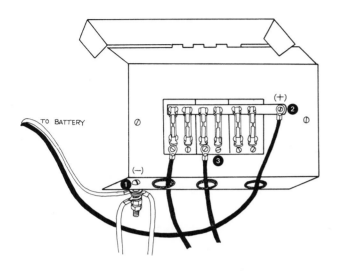

The positive (+) wire from your power source connects to the positive connection stud above the fuses (2). The positive (+) wires from the circuits in your house, usually grouped by the rooms they serve, attach respectively to the positive connection points below the fuses (3). Be sure to place 7/8-inch grommets in the holes in the bottom of the box before running wires through them.

When fuses are inserted between the positive wire connections, direct current will flow through them to the power paths emanating from the box—unless something is wrong in one of the circuits. Then the affected fuse will short out or "blow" to alert you to the problem.

With the addition of monitoring devices, your fuse box looks like this.

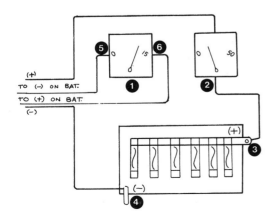

A voltmeter (1) and ammeter (2) will give you further checks on your system. Their wiring is directly related to your fuse box installation. The voltmeter, or battery-monitor device, shows the condition of the charge remaining in your battery. The ammeter shows how much current you are drawing at any one time through the fuse-box circuitry only. For instance, it will not show the amperage being drawn from an inverter or a water pump connected directly to a battery.

These meters are small luxuries that consume little energy to run themselves. You may choose to install just a simple battery-condition monitor with colored stripes (green, yellow, red) to indicate how much charge is left.

If you want an ammeter to monitor your current flow, you simply splice it into the positive wire coming into the power center from the power source, car plug, or inverter. Cut the wire just before it reaches the fuse box, connect the ammeter (2), then continue the wire to the positive connection stud above the fuses (3).

To install either the simplified condition monitor or the fully calibrated voltmeter, use a length of #22 white stranded wire. Solder or crimp on the appropriate wire terminals. Connect one end to the negative (−) stud on the meter (5) and the other end to the negative connection on the battery terminal. Then take a length of black #22 wire, add the terminals, and connect it between the meter's positive (+) stud (6) and the positive (+) battery terminal.

My voltmeter once shared the wiring to the fuse box, but I found that the reading was not as accurate as going directly to the battery because of voltage drop. Of course, if you have only the car system with no auxiliary batteries at your house, you must connect to the fuse box. To do so, connect the black positive (+) wire from the meter (6) to the fuse box at point (3) and connect the white negative (−) wire from point (5) on the meter to (4) on the fuse box.

Your fuse box and other components at the power center can be mounted with wood screws. Then you can cover the whole installation with cut-out and stained plywood or any other decorative face you have designed.

You have completed all the outside wiring to the house and, in fact, have done your first in-house wiring.

CHAPTER

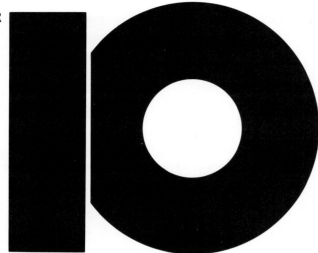

10

Wiring Your Home

At this point, you probably feel you have learned a lot about making wire connections, and your volt/ohm meter should have given you the hi-sign all along the path from the charging source to the house. But you still have not plugged anything in. Now you are going to put all the parts of the system together, and with your experience so far, it will be a snap.

Well, maybe not quite a snap. Get out your floor plans and assemble all the rolls of wire, tools, outlets, lights, switches, wire terminals, and connectors you have chosen to use. Take a look at the drawing of a complete installation on the next page.

As you can see, it really is fairly simple. You have all the pieces in front of you. Let us put them together.

One thing to notice in this view of the complete system is the inverter and 110-120-volt wiring. They are completely separate for reasons you already know by now. The installation instructions for the inverter are also separate and are covered in the following chapter.

In this chapter, I will explain just two simple 12-volt wiring processes, but I will offer you several alternatives along the way. You will perform the two processes over and over in different combinations, and then your wilderness home will be ready to light up.

One simple process is making the wire connections to the fixtures (outlets, switches, lights) you have chosen. The other process is wiring the power path along the wall, floor, or ceiling, between the installed fixture and your power center. You will be performing the two processes more or less simultaneously as you work, but for a clearer explanation here, I want to cover them separately. Let me start with the power paths.

Another name for the power path is the electrical circuit. One complete circuit is the path your wiring takes from the positive connection in your fuse box out to all the lights or outlets connected in parallel, then via the negative connections back to the ground terminal in the fuse box. One fuse makes the final connection to complete this one complete circuit.

For safety and convenience, each room in a house usually has at least two circuits. In that way, if you must someday do some trouble shooting along one circuit, you can still depend on the other one for light to see what you are doing.

There are two common approaches to this type of twin circuitry. For each room, you may connect all your built-in light fixtures along one circuit and all your 12-volt

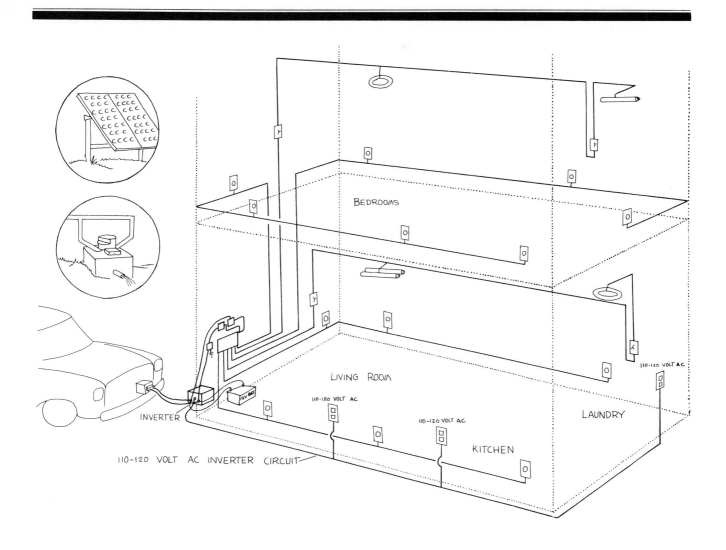

BEDROOMS

LIVING ROOM

INVERTER

110-120 VOLT AC

110-120 VOLT AC

110-120 VOLT AC INVERTER CIRCUIT

KITCHEN

LAUNDRY

110-120 VOLT A.C.

outlets along a second, separate circuit, or you may figuratively "split your room in half" and put both some of the outlets and some of the lights along each of two separate power paths. A separate circuit for communications equipment, stereo or TV, should not include fluorescent lighting fixtures that cause interference. In the illustrations for wiring to your fixtures later in this chapter, I will show details for both kinds of circuitry.

Once you have decided how to define your circuits for each room, you need to know where to put wire. There are three ways to install it: inside the walls, inside a specially made functional molding, or under the floor.

If you are just now building your home or have yet to

begin the installation, wiring inside the walls will be relatively simple. Assuming that your construction is traditional 2-inch by 4-inch framing, you can mount your switch and outlet boxes right on the wall studs. Then simply wire up your circuits and secure the run with nylon wire ties. Do not use staples to secure your wiring. They will often penetrate the wire and cause possible short circuits.

If your house will be made of concrete or logs, or perhaps will be one of those prefabricated domes of fiberglass or treated cardboard, an installation under the floor or inside the special wire molding may be preferable.

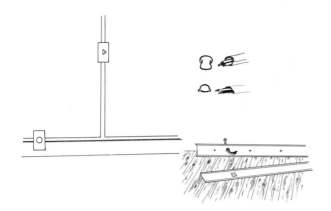

When you decide to use conduit, your outlets will have to be mounted in outlet boxes, and the conduit will be attached to the floor studs with standard U-brackets. But even so, the subfloor installation can be a great convenience when your home is an existing structure with nice, solid walls already in place.

Now I want to show you some sample connections along your power path. This first illustration shows the wire end of a 12-volt outlet under the floor.

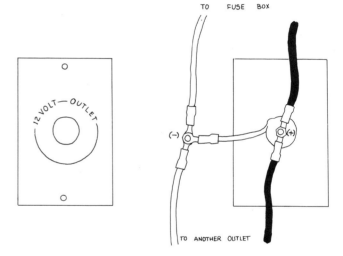

This wire molding is easy to install. You mount it along the baseboard and up the wall close to door or window frames. Electrical contractors carry a wide variety of styles and systems. They should be happy to help you choose the kind you want and plan the various lengths and corners you need. The cost will be about $7.50 for each ten-foot length. Look for an electrical supply house in the Yellow Pages of any good-sized town. Take a 12-volt outlet with you when you visit the dealer to make sure it will fit the available wire mold without modification.

If you must apply for a building permit and you say you will install a 12-volt powersystem, the officials may suggest running your wire through conduit to make it code acceptable. Conduit is also a practical necessity when you choose to run your power path under the floor.

I did not bother with conduit under my floor at first. For some reason, the little animals who take refuge under my wilderness home at night or during storms were attracted to my wiring. When they nibbled on it, my lights went out. Of course, I located the problem very quickly by following the circuit with the blown fuse. I have encased all my sub-floor wiring in conduit now. The system is easier on fuses and on the little animals as well.

If you have access under the floor, the 12-volt outlet can be installed very quickly. Just drill a 1 1/8-inch hole with your brace and bit. Then push the wire end of the receptacle through the hole and connect it. Make sure that you measure for the hole so you drill through just the floor, missing any supports beneath it.

This metal outlet, rated at 10 amps or less, has a chrome face plate. On the back, a screw and nut make the positive wire connection, and a crimp-on terminal makes the negative wire connection. This type of outlet, common in most RV and marine applications, serves as the model for the rest of this chapter.

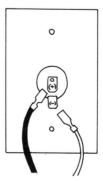

This newer model, rated at 15 amps, has metal contact areas encased in hard plastic. Both positive and negative connections are made with slip-on connectors crimped to the wires. I have found this model easier to install, and recommend that you look for it in new installations.

On the back of either outlet, one pair of black positive (+) and white negative (−) wires come in from the fuse box. Then on the other end, the pair of wires runs along the underside of the floor to another outlet, secured to framing with nylon wire ties. Note the regular ring terminals overlapping each other for the positive (+) connection on the back of the metal outlet itself and the three-way terminal that connects the negative (−) pole of the metal outlet with the ground wire running alongside. In the actual installation, you would cover the exposed terminals with sufficient electrician's tape for insulation.

If you are still building and have access inside the walls, the same outlet can be connected with the same wire terminals as illustrated below.

This time, you place your installation inside a regular wall box, which is attached with nails to the side of a stud in the exposed wall. The top pair of wires continues up the stud to a light fixture or switch, and the bottom pair runs back to the fuse box.

Those are the simplest, most basic wiring-to-fixture installations, but they are not necessarily the best. No conduit is involved, and the time spent soldering terminals is not always necessary.

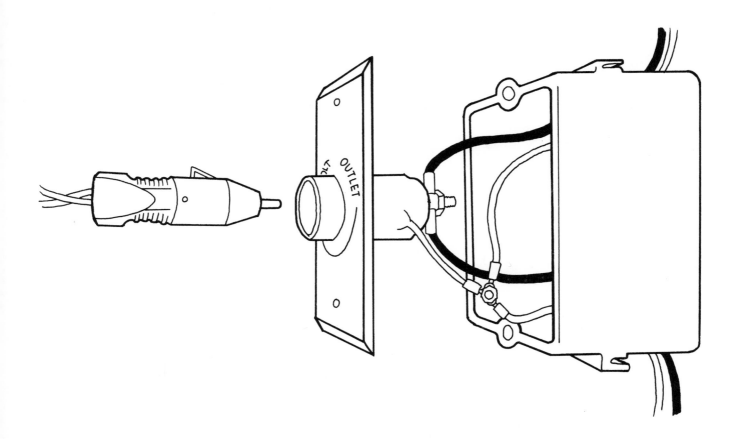

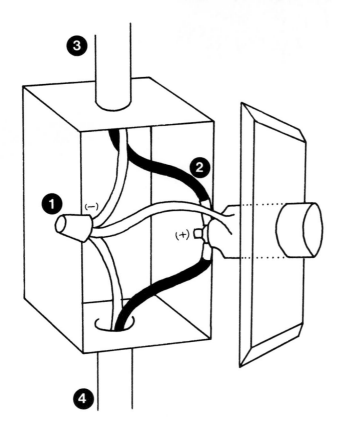

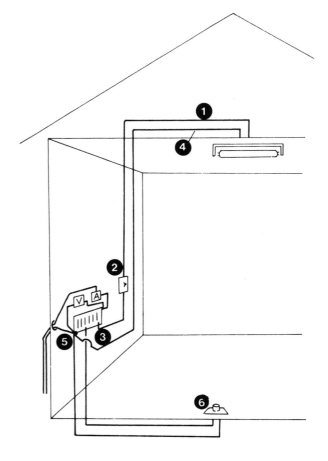

again. Remember to connect black wires to black and white wires to white to preserve your positive (+) and negative (−) polarity. And remember to run at least two circuits to each room for better control later.

Here is a modified installation of the same outlet. A closed-end nylon wire terminal (1), also known as a wire nut, is used to connect the negative (−) wires. These wire nuts are available both with and without coiled springs inside them. Be sure you get the type with a coiled spring to insure a good, tight connection. The positive (+) connection is again made with ring terminals (2) as in the previous illustrations.

Notice how our 12-volt outlet fits into the regular outlet box without modification. The box permits ready hook-up to the electrical conduit that runs to a wall switch (3) and back to the fuse box (4).

The way you choose to wire a light, outlet, or switch will often depend upon the location of that unit in relation to the wires that lead to other units, and the wires that return to the fuse box. As we turn to the connections with your fixtures now, I will rely on diagrams that show a wide variety of procedures for installations.

Two things to keep always in mind bear repeating

In our basic system, several simple installations are obvious. A ceiling-mounted light fixture is connected by black wire (1) to the fixture's black wire with a wire nut, then connected to a wall switch (2) with #10-12 hook terminals. The black wire then continues to one of the positive terminals in the fuse box (3).

The white ground wire (4) is connected to the light's ground wire the same way, with a wire nut, then alongside the switch all the way back to the ground terminal at the fuse box (5).

Note that this cutaway view also lets us see the side of a 12-volt outlet (6) in its subfloor installation. The wiring can be virtually the same as that for the light fixture, except there is no switch. You may also use the regular and three-way ring terminals, which I described earlier.

For the rest of the chapter, I will show you several combinations of installation alternatives. I have tried to anticipate the component procedures for a wide variety of combinations you may devise. The first example includes two different illustrations—a schematic-type wiring diagram and a more realistic detail drawing. Then we will follow through with the realistic drawings.

The most basic wiring is the connection of a single 12-volt outlet. In the schematic drawing on the left, you can see how the single outlet relates to the whole 12-volt battery system you have installed so far. Note that the fuse box is literally at the center of the system and that the integrity of black positive (+) wiring and white negative (−) wiring is preserved on both sides.

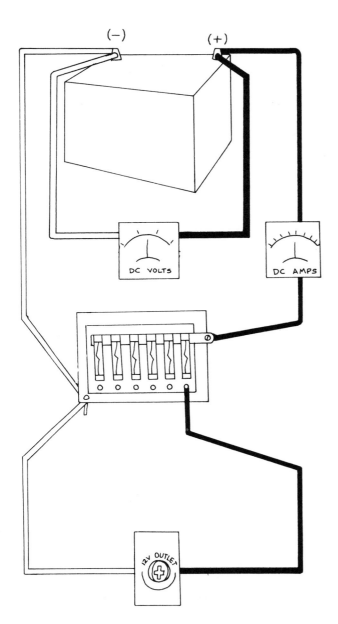

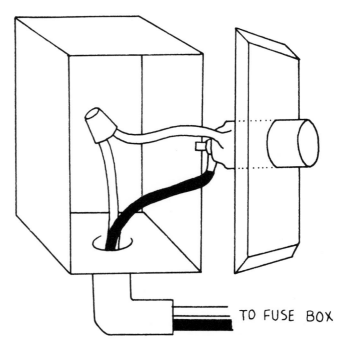

In the more realistic detail above, you can see the black wire connected to the back of the outlet with a ring terminal, the white wires crimped together in a wire nut, and both wires returning to the fuse box.

Here is the same simple outlet, but it is no longer "hot" all the time. Now you have a wall switch in the circuit, so you can turn it on and off from another part of the room. The wiring is still simple, with wire terminals on the black wires and wire nuts for the white in both the switch and the outlet.

Your first light fixture is installed in another simple variation on the single outlet and switch wiring, but notice that this switch is at the end of the wiring, rather than between the fixture and the fuse box. The light fixture itself is connected with wire nuts on both black and white wires.

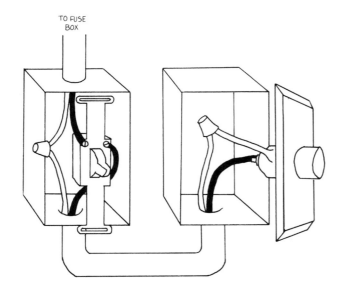

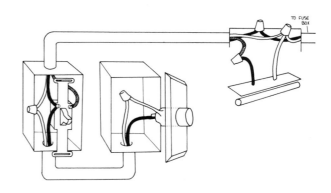

In the following variation on the first connection, the switch is eliminated, but you have a second hot outlet on the same circuit. Obviously, the switch could be restored, no matter how many outlets are in the parallel circuit.

The next installation is a single circuit, but it includes both a hot outlet and a light fixture with a wall switch. Note the addition of a third wire color, usually red (striped in our black-and-white drawing). The red wire is used for the positive path from the switch to the light, distinguishing it from the other positive wire to the outlet when both are in the same circuit. The red-and-black wire distinction can prevent much confusion, both during your installation and in routine service or troubleshooting years from now.

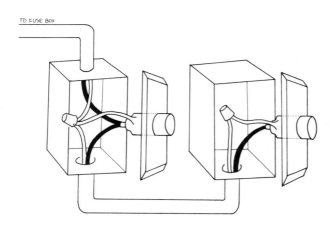

The result of the following installation in your home will appear to be exactly like the installation we just saw—one hot outlet and one light fixture with a wall switch—but this time, the light and the outlet are each on their own circuits. If one fuse should blow, you can trouble shoot that circuit by the light of the other circuit.

This installation is really a combination of the very first hot outlet and the first light fixture and switch you saw earlier. With two separate circuits, we have chosen not to use the red wire.

You can also expand the circuit with the light and switch. In this single circuit, you have two lights that operate from a single switch.

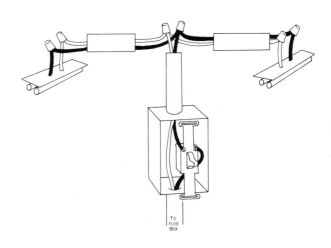

Another alternative for the same light-fixture circuit offers you two three-way switches that control one light fixture, presumably from opposite ends of a large room or at the top and bottom of a stairwell.

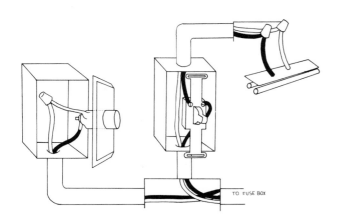

Here is an expansion of the two-circuit wiring in which we have connected on one path a series of hot outlets that do not affect the simple light and switch on the other.

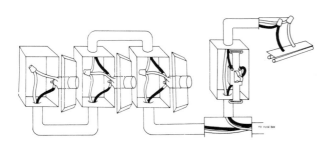

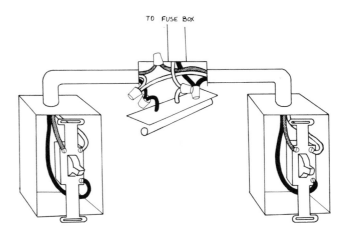

This final variation gives you two lights, each with its own switch, on one circuit. The left light fixture is operated by the switch on the left, the right fixture by the right switch, with both switches in the same wall box. All wires in the box are positive but of different colors to avoid confusion.

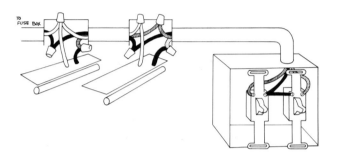

Even if you are brand new to home wiring, you can see from these examples that the installation is nothing more than a sequence of logical combinations. The possible alternative combinations are practically unlimited, and only you can put them together to serve your own needs best.

Take the time to plan for your own comfort and convenience, then wire together the combinations of outlets and fixtures, with or without switches, that will meet your requirements. Just keep in mind the completion of each circuit back to the fuse box and the distinctions, color-coded when necessary, between wires for each device installed along one power path.

If you have chosen to install a power inverter, go on to the next chapter. If you want my help with a water system, go on to the chapter after that. But if you do not need the water or inverter, you have completed your installation. Skip ahead to Chapter 13 for a checklist and some last-minute precautions. Then you can reconnect those battery terminals you left undone for safety, and get set to light up.

Your Inverter System

Following the procedures in the last chapter, you equipped your wilderness home with 12-volt power in a series of electrical circuits. Now, if you have chosen to include a power inverter in your total powersystem, you will add one more distinct circuit. This power path will carry 110-120 volts of alternating current, completely separate from the 12-volt direct-current wiring, except at the power source.

We will deal here with just one simple circuit that meets a maximum demand of 15 amps from your system. For very high AC demands, you really should look at a generator rather than an inverter working from your 12-volt system, but you should have resolved this question as you designed your system in Part One. None of the inverters listed there will power more than two AC circuits, and only one circuit is required to energize the maximum number of outlets and the minimum number of AC accessories in most of our wilderness homes.

As you work on the wiring for your inverter, please keep in mind that the AC circuit is very different from all the low-voltage circuits with their inherent relative safety. You must observe all the cautions that relate to 110-120-volt AC electricity, because that is exactly what you are handling. If you should be careless with this installation, no slight tickle can jar you back to caution—rather, a large, unpleasant shock will be the result.

Check the requirements of the building code in your area. You may want to hire a qualified electrician to help if your plans call for extensive AC wiring.

Although these extra precautions are absolutely necessary, I certainly do not mean to discourage you from the inverter installation. The added convenience and luxuries permitted by my inverter system at home are well worth the extra care I exercised during its installation.

Your power inverter will last indefinitely if you use it within its rated capacity. If you overload the inverter—ask it to supply more than its rated wattage output—you will burn out the transistors that make it work. Check the wattage on the model you have selected. The wattage or amperage limits usually appear on the name plate. (The symbol VA on the name plate means watts.) If only amperage appears, multiply that figure by 110 or 120 volts to get approximate wattage. Then be sure that the wattage of any accessory or small appliance you connect is smaller than the rated wattage output of your inverter.

As an alternative to a large, permanent inverter installation, or as a supplement, there are now a number of small plug-in inverters that can operate in conjunction with a particular small appliance. The small inverters are energy savers because they avoid the need to activate the big unit to run small appliances, but be sure to observe the difference between continuous and motor-starting wattage.

As an example, you may choose a little 200-watt inverter attached to a portable sewing machine. You can operate both from any room in the house equipped with 12-volt outlets. Simply insert the inverter's cigarette-lighter-type plug in the outlet, and use an ordinary household extension cord between the sewing machine and the inverter. Although the manufacturer's rating on most home sewing machines is below 100 watts, I still recommend the little 200-watt inverter. Like many small appliances, these machines have motor-starting requirements higher than the 100-watt rating for continuous operation.

Now, hopefully somewhere between these considerations and the design you created earlier, you have chosen the right power inverter for your needs, and you are anxious to get it installed. Here is a quick list of the other parts and materials you need to have on hand for this special wiring job. Check the illustrations on the following pages as you go through the list, and you will begin to see how it all fits together.

The wire itself for your 110-120-volt system, to be run between the inverter AC outlet and the wall outlets in the house, can be Romex—those three solid wires encased in tough plastic that we discussed earlier. Romex is the safest wire you can use here, and you should get the #12/3-600 volt size. The length you need is determined by the distance from the inverter to your AC outlets. You may connect as many outlets as you need in parallel on a single circuit, throughout the house. Since the three wires are encased together, you do not need to measure the return length as you did for the DC system. You can buy Romex at a building supply or hardware store for about 20 cents per foot.

You will need a standard 3-pronged plug on the end of the Romex wiring to connect your AC circuit to the 110-120-volt outlet on the inverter. Be sure the plug is rated for at least 15 amps. Heavy-duty plugs are available at building supply or hardware stores for about $1.50.

Another solenoid is required to isolate the inverter from your 12-volt system and activate it from a remote location. This solenoid can be identical to the battery isolator you installed with the auxiliary battery in the car—rated for 80 amps or more, continuous duty, normally open position. You can buy it for about $10 at an electronics, RV-camper, or marine supply store.

You will need some white #22 stranded wire and various terminals to fit. This wire will be connected to your new solenoid and run between your wall switches in the house to activate your inverter. To find the amount of wire you need, measure the distance from the inverter to the switches and from the switches to the nearest 12-volt white ground wire for a tap. This is the only exception to the rule not to tap into a nearby ground wire.

You will need a two-inch piece of #22 black wire and terminals to fashion a jumper, just like you used to wire your car. The jumper makes a positive wire loop to complete the remote circuit on the new solenoid. The #22 wire costs about eight cents a foot.

Many inverters are equipped with red pilot lights which signal that they are on and drawing amperage from your system. But if your actual inverter installation is out of sight in a battery box, you may want to install another pilot light right at the remote switch or outlet from which you operate the inverter. Be sure you get a light rated for 110-120-volts AC. The cost at an electronics supply store is about $2.

Your design will call for one or more regular AC wall switches and outlets, with mounting boxes and cover plates. I recommend the convenient combined switch and outlet in the illustration, requiring a standard double wall box. Each of these combinations will cost about $3 at a building supply or hardware store.

The first step in the installation is to pick the best location. To avoid voltage drop, your inverter should be placed as close to your batteries as possible—no more than six feet away. With auxiliary batteries on-site, the location is easy—inside your home at the power center or outside in a special box. You should be aware of the inverter's need for some ventilation, and locate it as far as possible from any other heat sources. Do not block the flow of air around the inverter, or it may overheat and burn out.

Page 115 shows a typical installation that assumes the home operates with an auxiliary battery in the house. Notice that the house-to-car plug can serve as a backup source or as the main power source, assuming you have

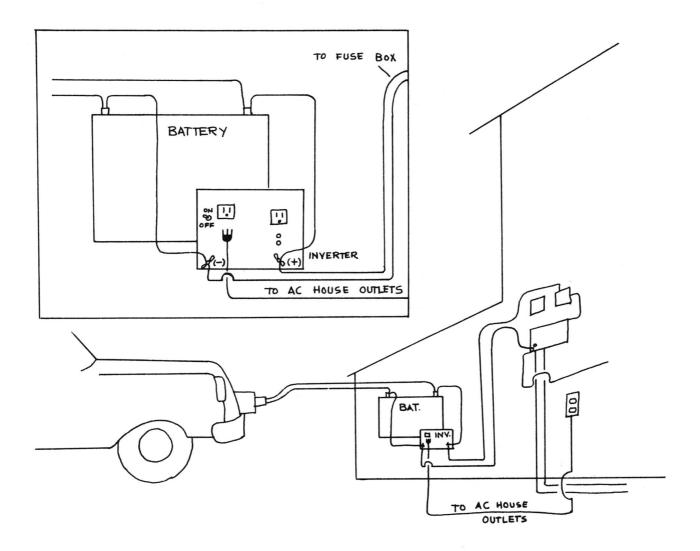

installed a dual-battery switch. Whatever the power source, it should be connected first to your inverter, then to the fuse box. In this order, you gain the inverter's maximum AC voltage output from your battery's maximum 12-volt output. Make your connection directly to the power source, with the heaviest wire available, and be sure to observe battery polarity.

Let us review the kinds and types of wire needed from the power source, based on the size of the inverter.

250 watt or less—#8 stranded automotive

251 watt to 400 watt—#6 stranded automotive

401 watt to 600 watt—#4 stranded automotive

601 watt to 1,000 watt—#2 stranded

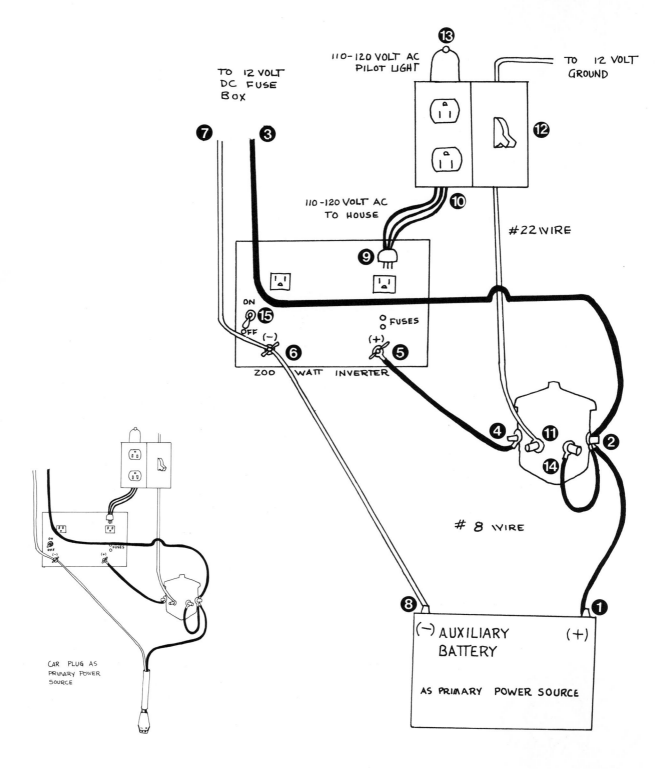

TO 12 VOLT
DC FUSE
BOX

110-120 VOLT AC
PILOT LIGHT

TO 12 VOLT
GROUND

❼ ❸

❶❸

❶❷

110-120 VOLT AC
TO HOUSE

❿

#22 WIRE

❾

ON
❶❺
OFF (−)

❻

o FUSES

(+)
❺

200 WATT INVERTER

❹ ❶❶

❷

❶❹

8 WIRE

❽

❶

(−) AUXILIARY   (+)
BATTERY

AS PRIMARY POWER SOURCE

CAR PLUG AS
PRIMARY POWER
SOURCE

These diagrams contain all the steps for your complete inverter installation. In the inset, note how the car plug is situated when being used as the primary power source. Used as a backup system, the wires from the plug connect to the auxiliary battery terminals, then to the inverter, then to a dual-battery switch (not shown), then to the fuse box. In either case, the black positive (+) wire passes through the solenoid only if remote control of the inverter is to be installed.

We will begin our installation at the battery, but be sure you do not complete the battery connection until everything else is in place and all switches are in the "off" position. Do not take any chances with this tremendous voltage boost as you add it to your 12-volt system.

Take a length of #8 (or heavier) black wire to run from the positive (+) post of the battery (1) to one side of the new solenoid (2). Solder on wire terminals and complete the connection on each end.

Take another length of #8 (or heavier) black wire, attach wire terminals, and make the connection from the solenoid (2) to the positive (+) connection on the fuse box (3).

Take a short length of #8 (or heavier) black wire to reach from the other side of the solenoid (4) to the positive connection on the power inverter (5). Solder on wire terminals and complete the connection.

Take a piece of #8 (or heavier) white wire to run from the negative (−) post on the inverter (6) to the negative (−) connection on the fuse box (7). Attach wire terminals and make the connections.

Cut another length of #8 (or heavier) white wire, attach the appropriate terminals, and make one connection to the negative terminal (−) on the inverter (6). Leave the other end of this wire disconnected from the battery (8). For safety reasons, this connection will be the very last step in the procedure.

Now your placement of the inverter at the power source is complete except for two small connections on the solenoid. If you have installed your inverter right inside the house, without a solenoid for remote control, you can plug your small AC appliances directly into the 110–120-volt AC outlets on the inverter itself. On the other hand, in a larger home, you may want to install several outlets equipped with switches to operate the inverter from several convenient locations. We will come back to the last two connections after your switches are installed.

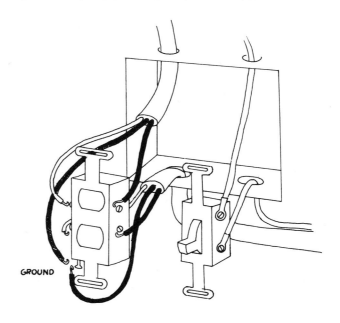

GROUND

To begin, you must mount wall boxes that will contain the switch and outlet combination at the locations you have chosen. Do not put in the switch or outlet itself until the wiring is in place.

Start at the inverter, strip away the plastic covering on the Romex with a Romex tool. You will see that there are three wires, one with black insulation, one with white insulation, and one that is bare copper. Strip away 1/2 inch of the insulation. The integrity of connecting black wire to black and white to white is as important here as with your DC system, though no fuse will blow and it will not affect the inverter or AC appliance operation if you inadvertently cross the AC wires. Keeping the wire color continuity in an AC system allows for easy tracing of the wiring if service is ever required. In many tools or appliances, a shock hazard will exist if the color coding is not followed religiously.

Now return to the numbered drawing for the overall inverter installation. Connect the three wires from your outlet to the inverter plug (9) by bending the wire ends into hooks, making sure that the middle copper wire is connected to the green ground-connection screw in the plug. Do not plug into the inverter yet. Measure the distance to the first wall box (10). Cut and run a length of Romex to it. Make sure it is long enough to leave about six inches protruding from the box. Do the same thing from the first wall box to the second, third, and so on.

You do exactly the same with the #22 white wire as with the Romex, except that you run lengths of it from the inverter solenoid terminal (11), alongside the Romex, to the wall box on the switch side (12). Again, leave at least six inches of wire protruding from the box.

Each wall box with a switch requires a #22 wire connection from the solenoid. Another white #22 wire is required to run from the first switch and wall box and continue on to the second, third, and so on, as shown. Running your wiring to each switch this way will allow any one of them to operate your inverter.

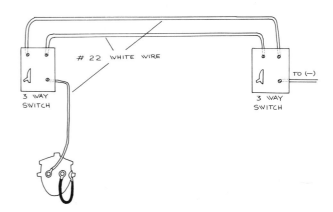

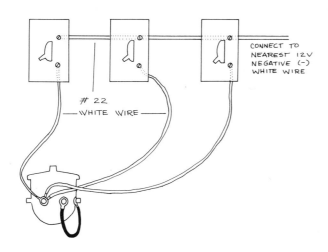

When you reach the last box, simply connect the three Romex wires to the AC outlet. For the switches, continue on with the white #22 wire to the nearest white 12-volt ground wire and tap into it with a three-way terminal. Leave the cover plates off the boxes until you have checked out the system later.

There is one inconvenience with wiring the three or more switches in parallel, as we have just done. You can begin the inverter's operation from any one of the switches, but to turn it off, you must return to the same switch. If you want the convenience of turning on the inverter at one switch and turning it off at another, you can install two three-way switches.

Your convenient three-way switches can be wired as shown. Note, however, that more than two three-way switches cannot operate the same fixture, as is true in any wiring situation.

You may wish to install a 110–120-volt pilot light above one or more of the outlets. The fixture illustrated here will signal that the inverter is operating.

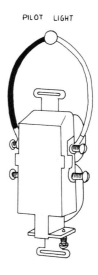

The same light is pictured on our overall installation (13) on page 116. It connects easily to the back of the AC outlet, right beside the Romex wires you connected. Black and white #22 wire is all you need to make the connection between the light and outlet.

Continuing with the installation on page 116, you are ready to make the last necessary connections. Take the two-inch #22 black jumper wire you fashioned and connect it to form a loop between points (2) and (14) of the solenoid. This completes the 12-volt circuit to operate the solenoid when a wall switch is turned on.

Now you are ready to test your AC system.

- Insert the plug (9) into an AC outlet on the inverter.
- Make sure the main inverter switch is off (15).
- Check the fuses on the inverter to make sure they are good.
- Connect the power source (8) you left undone when you began the installation.
- Flip a wall switch to the "on" position.
- Turn on the main inverter switch.

You should now have 110-120-volts AC, indicated by a glowing pilot light at the outlet. Try out an appliance. If your inverter does not come on, recheck your connections and installation procedures. When everything checks out, go ahead and finish the installation by placing the covers on your outlet and switch boxes. Leave the inverter's main switch in the "on" position. Your remote switches in the house will actually determine whether it is on or off.

The power inverter is a major feature of your total powersystem. You will be grateful for the extra comfort and convenience it allows. But you also need to be aware of some special situations involved.

A slight 60-cycle hum in your inverter is normal. A rattle or buzz is due to either loose transformer laminations or a loose transformer coil. A small wedge of wood or paper driven between the transformer coil and the iron laminations will often stop it. Electrician's tape around the iron laminations can also help.

A small black line appearing at the top or bottom of your TV picture is normal and is not a defect in either the TV or the inverter. It will not hurt either. A slight adjustment of the vertical hold will often correct the situation.

Some new TV sets have been equipped with special power supplies designed to protect against low-voltage conditions. This feature interprets the square-wave inverter's output as low voltage. A neighbor of mine tried to run one of these sets, rated at only 120 watts, on a 200-watt square-wave inverter. His tuning dials were lighted and, inside the set, the back of his picture tube showed illumination, but he was receiving neither picture nor volume. We tested his set on my 550-watt inverter and got the same result. The solution was to take the set and the schematic portion of its instruction manual to a qualified TV technician, who modified the power-supply feature. This solution did void the warranty on the set. One alternative was to get a rotary inverter with sine-wave output, but perhaps the best solution would have been to sell the set and get a 12-volt TV.

If your radio or amplifier makes more noise than usual, it probably needs a better power-supply filter. Certain inexpensive equipment will make noise with the inverter. Higher quality equipment usually will not. A radio or TV repair shop can install additional filtering in your radio or amplifier. You may ask a radio or TV dealer to install a hash filter in your inverter.

Some test equipment and certain brushless motors (synchronous, shaded pole, hysteresis) may not work well on the inverter's square-wave power. Certain small appliances within the power rating of a square-wave inverter require from three to seven times their rated wattage to start. Trying to operate them will result in inverter failure. As a general rule, brush-driven motors are okay and brushless are not. If you have any doubt, check with the motor manufacturer or your appliance dealer.

I mentioned earlier that some inverters have a red neon pilot light to signal that they are on. The light is also a safety feature. When the light is on, the battery/inverter/appliance chain is functioning normally. If the pilot light should flicker or go off while in use, your appliance is overloading the inverter, or your battery voltage is getting too low. If your battery voltage is too low to begin with, the inverter will not come on at all.

Your inverter will give more reliable service if you turn off or disconnect the appliance before turning off the inverter.

If your phonograph and tape recorder seem to play too fast or too slowly on the inverter, that is a special problem. Only the more expensive frequency-controlled inverters will make your phonograph, tape recorder, and other frequency-sensitive devices run normally. The best solution is to get 12-volt equipment. I sold my AC stereo system for enough to buy a 12-volt DC system without additional cost.

If you have chosen the sine-wave operation of a rotary inverter instead of the square-wave model, you can avoid some of the special wiring connections for remote operation. Every rotary model is equipped with a solid-state device to sense load demand and turn on automatically. Obviously, a pilot light is unnecessary, and in fact, would appear as a demand on the sensor of a rotary inverter. Like the solid-state inverter, the rotary unit should be located close to your batteries. The connections are simple, as shown here.

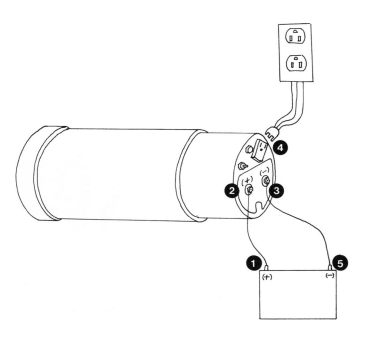

To install your rotary inverter, select a length of #8 (or heavier) black wire to reach from the positive (+)

battery terminal (1) to the positive (+) connection on the rotary inverter (2). Solder on the appropriate terminals and make the connection.

Take a matching length of #8 (or heavier) white wire, attach the appropriate terminal, and connect one end to the negative (−) connection on the rotary inverter (3). Leave the other end disconnected for safety.

Attach a plug to the end of the Romex wires leading from the 110-120-volt AC circuit installed in your house (4), according to the instructions for a solid-state inverter installation. Insert the plug into the AC outlet on the rotary inverter (4).

Make the final connection by attaching the white wire to the negative (−) terminal on the rotary inverter (5).

Be careful about the location you choose for the rotary inverter. The automatic demand-sensing is affected by moisture, which can start inverter operation for no apparent reason and keep it running even after all appliances are turned off. So choose a cool, dry place to install your rotary unit. You may also want to consider interrupting the 12-volt wiring at a switch you can turn off when leaving home. Doing so will cut off power from the battery system and provide insurance against false starts while you are away.

That is it for your power inverter installation. If you have decided on a 12-volt water system, let us move on to that. If not, you can move ahead to Chapter 13.

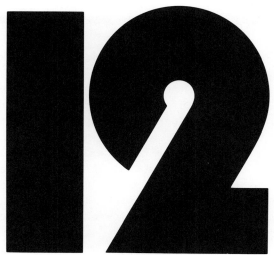

CHAPTER

12

Your Water-Pumping System

Someone else can write a whole book about all the alternatives to consider when you try to bring water to a wilderness home. I wish I could have read that book when I began to create my combination 12-volt water system by trial and error. Instead, I had to learn from experience until I finally put together the one simple system that works best for me.

One traditional power source to pump water into a holding tank from a long distance is the good old portable AC generator. For either deep or shallow wells, there are a number of good 110-120-volt AC electric pumps. Some models are submersible right into your water supply. To use an electric pump, you will, of course, need a traditional generator. With a generator of sufficient size, you can operate a high-amperage battery charger while pumping water, socking away 80 or more amps in your battery at the same time. So, although you may object to the noise of a gas generator, minimal use with the charger makes good sense.

Then there are strong, slightly noisy, pumps driven by gasoline engines. I use one periodically to siphon water from a well and pump it to the holding tank. When the level in the tank gets full, my quiet little 12-volt pump takes over to provide water pressure from the holding tank to my home as needed.

A windmill is an even older tradition and a good way to drive yet a different kind of pump to bring water to your holding tank. Then your 12-volt system can pump the water from the tank to your home on demand. But you must be located in the right geographic location to use the wind.

My home's water supply depends on a 3,000-gallon redwood holding tank that sits about 100 feet from the house. I pump water into the tank from a well by conventional means during the dry summer months. In the winter, water flows into the tank through a pipe located in a stream several hundred feet away and only slightly higher than the tank. I do not have any water under continuous high pressure to the house at any time. I use a 12-volt pump to get the necessary pressure. The good thing about this system is that I never fear a pipe bursting when I am not around. When the car is not plugged in, or I leave and shut down my auxiliary battery system, there is little pressure in the pipes.

Anyway, if you do not have to draw water from more than a ten-foot depth and push it more than 25 feet at the same time, this 12-volt water pump will work well for you. If your requirements exceed these demands, you will need a more powerful pumping system and should seek expert advice.

Obviously, the 12-volt pump can be useful in a variety of situations. The problem then is to find a 12-volt pump that will withstand constant use in a year-round residence. I tried several makes that lasted three to six months, some for even shorter periods of time. Many are built cheaply in anticipation of only infrequent use in RVs during the summer or other vacation times.

I have found two very reliable 12-volt pumps—the Jabsco 3800 series and the Coleman Aquarius model 2248-730. The Jabsco model is used for the system illustrated in this chapter.

The good pumps are shipped with full and easy installation instructions, as well as comprehensive troubleshooting charts. My limited suggestions here are just to provide additional guidance. Be sure you follow the manufacturer's recommendations, especially when they affect the warranty.

Once you have figured how to carry the water to a holding tank and chosen a reliable 12-volt pump, you will need an override switch and a fuse block, or an in-line fuse of 10 amps. The other materials you will need are #10 black and white stranded wire, pipe, joints or clamps, pipe size-reducers, and shut-off valves. The actual sizes will vary with your installation, and so will the total cost. It should be relatively inexpensive.

My water system, pictured here, includes 1 1/4-inch PVC pipe placed in an 18-inch-deep trench between my holding tank, which is level with the house, and the 12-volt pump. Note that the pump is installed in a box. I have shown the battery and other electrical components in the box so you will get a clear picture of the system.

One precaution is the installation of shut-off valves at both ends of the pipe, (1) and (2). If anything should ever strike the buried pipe, causing a leak, I can preserve my stored water by shutting off the pipe right at the tank (1). I use the other valve (2) whenever I am working on the pump.

The pump inlet itself is only 1/2 inch in diameter. You can connect the main PVC pipe with a series of pipe reducers (3) and a short length of the smaller-size pipe. The pump outlet (4) also takes 1/2-inch pipe, and you may need a 1/2-inch by 3/4-inch pipe reducer if the cold water main in your existing house is 3/4 inch in diameter, as most are. If you are installing your own main pipe, the 1/2 inch diameter is too restrictive.

Rather than using the different PVC pipe sizes from point (3) on, I used potable hosing, which is approved for drinking water, and metal clamps to make secure connections to the pump at (3) and (4). The water pressure in the 12-volt system is much lower than the water pressure that most building codes were designed to cover, so to my mind, the convenience of the hose is a legitimate exception, but it will not be if you are seeking code approval. Removal of my pump for any reason is much easier with the hose.

If you decide to use PVC pipe and make permanent connections all the way, install universal pipe joints near points (5) and (6), and you will gain relatively easy removal of your pump. You might take this illustration, along with the instructions that came with your pump, to the people in your building supply or hardware store. Then they can help you find exactly the pipe and connections you need to fit variations in your particular installation. While you are there, be sure to ask for their recommendations on glue or adhesive for PVC pipe connections.

In smaller water systems like mine, a problem sometimes occurs with restricted space in the water lines. When you turn on an inside faucet just halfway or less, or when you flush a toilet, you may see rapid on-off cycling by the pump. This is hard on the pump and causes an unnecessary waste of power. To help alleviate this problem, install an air-accumulator tank in the water line, somewhere between the pump outlet (4) and the cold-water main (7). The accumulator tank, illustrated here, will cost about $10 at RV-camper or marine supply stores.

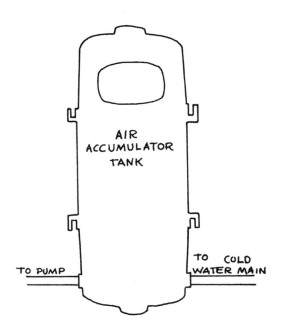

When you have completed the installation of your pump, you will have to wire it. If your power source is the car, disconnect the powering battery. Follow the manufacturer's recommendations for the pump wiring, which, in every case I have seen, match the illustration. There are two main wire connections to be made directly to the battery, not to the fuse box. You want to keep your pump motor from affecting your lighting circuits as much as possible.

Measure and cut a length of #10 black positive (+) wire and connect it at point (8), then to the fuse at point (9). Cut another length and connect it to an override switch at your power center (10). This will allow you to turn the pump off quickly from the inside if necessary. Keep the switch in the "off" position. Continue with another length of wire to the positive (+) terminal of the battery (11) and make the connection. Cut a length of #10 white wire to reach from the negative (−) wire on the pump (12) to the negative (−) battery terminal (13). Secure it with the proper wire terminals.

If your home is not equipped with an auxiliary battery, use the copper bolts suggested for connecting wires in junction boxes, Chapter 9. Tap into the car plug wires along their run to the fuse box and make the pump connection. If you use the copper bolts and tap into the wiring, be sure you wrap the connections with plenty of electrician's tape for insulation. If an inverter has been installed, make the pump connection to its 12-volt terminals.

You must not fill in the trench for your pipe from the holding tank to the pump until you have checked the installation thoroughly to be sure there are no leaks and everything is working. When all is in place, make sure the override switch is in the off position. Go back to the disconnected battery in your car. Make the final connection you left open for safety.

Open up your shut-off valves, then go inside the house and turn on your override switch. Turn on a faucet at the sink to expel accumulated air. Then your pump should operate and produce water. If not, check the fuses first. Make sure you have not reversed the positive (+) and negative (−) wires, and check all your connections. Double-check your pipe connections for leaks. Then try again.

Once your water system is operating, fill in the trench. Then go back in the house, and enjoy the luxury of water.

Here are a few basic tips for the routine use and maintenance of your pump.

If you notice that your pump's motor is running for periods of time when no one is using water, turn it off inside at the override switch while you locate the problem. There usually will be a leak somewhere in the water lines. Check first for loose pipe connections in the house to be tightened or for worn gaskets to be replaced in faucets. When you get rid of the offending leak, the constant on-off cycling of the pump's motor will be reduced considerably. Some intermittent cycling is normal and keeps the pressure up in your lines.

The most consistent problem with even good, heavy-duty pumps is a broken impeller or belt. These parts are easily repaired, but living in the wilderness, you may want to keep a standby pump on hand for quick replacement while you determine the problem and solve it. Then there will be no interruption of water to the home. In any event, have spare parts handy for on-site repair.

Your pump has a powerful little motor, and its operation may create interference on your TV screen or in radio reception. Check with a radio-TV dealer or an auto radio dealer for a noise filter that reduces electric motor interference. It is easily installed and inexpensive.

As I review these considerations regarding my 12-volt water system now, they all sound so simple. I will not forget the months of frustration I endured to be able to boil it all down to this one short chapter, and I hope you can profit by my experience.

Your water system should be installed and operational now. If you chose a power inverter, it should be in place and ready to go. In fact, your entire 12-volt Wilderness Home Powersystem is now complete.

It Is Installed

Now you have progressed from the wiring in your car, through the space between various charging devices and your house, to the power paths and all the appliances and accessories on the system you designed for your wilderness home. You may have installed a power inverter for 110-120-volt AC electricity and a pump for automatic water pressure on demand. Before you turn on all your lights, radio, TV, food mixer, and blender, look back and check or double-check some of the main points that will permit you to enjoy living with your system.

Did you always connect white wires to white and black wires to black?

Have you been through the whole powersystem with your volt/ohm meter, checking for sudden voltage drops or weak connections?

Is every wire connection inside the car, outside the house, and in your home just as secure and tight as you can make it? Did you solder every connection you should have?

Is your powersocket or other receptacle installed on the front bumper of the car so that you will never risk pumping carbon monoxide into your home?

Are auxiliary batteries in your car securely bolted or strapped into place?

Is your battery box properly located, insulated, and ventilated?

Is all the wiring to junction boxes or under the floor secured and covered? Are all the car-to-house wire connections sealed? Is the whole outdoor setup weatherproofed and impervious to small animals and other visitors?

Did you install fuse boxes or fuse blocks as early warnings in every circuit in your car and your home?

Have you carefully followed manufacturer's instructions for the installation of alternative charging systems—or any other equipment protected by warranty?

Does every white negative (−) wire (except that for the inverter remote control) inside your home run all the way back to the ground-terminal bolt on the fuse box?

Is every power path along the walls of your home secured carefully with nylon wire ties, never stapled? Is the path through the wire molding good and snug?

When you rewired or added 12-volt plugs on some small fluorescent lamps or appliances, were you careful to observe the positive-negative polarity rule?

Have you chosen carefully a power inverter with the right capacity for your needs? Did you install wiring large enough to meet the needs of the inverter?

Did you install larger wiring to the kitchen or laundry areas for the larger appliances of the near future? To second or third stories of the house? Wherever your meter or figuring showed potential voltage drop?

Is your 110-120-volt AC circuit safe and installed according to the building codes?

Has your water system been double-checked for leaks, both inside and outside the house?

Did you get noise filters to reduce interference by your pump or other motors in the reception on your communications equipment?

If you were able to answer "yes" to all these questions, you have done an excellent installation. If you did a car installation, connect up the remaining wire to your auxiliary battery and plug your house into your car. If you installed alternative charging systems, make those final connections and double-check them.

Gather your family and friends together. Turn on a few lights, unpack your glasses, and get some crushed ice. Put your best dancing music on a 12-volt stereo. Turn every damned thing on. Switch on the inverter and run your blender for a minute.

Frozen daiquiris, anyone?

14

Living with Your System

Have you ever met a twelvoltaholic? You are reading the words of one right now. You will probably become one yourself in the near future. Twelvoltaholism is a condition that tends to become a pretty good way of life, and I am delighted when I get to meet some of the many others in the same condition today.

This condition is characterized by two kinds of behavior. First, we are avid readers of advertising and catalogs that introduce all the new 12-volt appliances and accessories. So many breakthroughs are announced every week that I have trouble not getting excited about installing them and experimenting with them.

The second aspect of my twelvoltaholic behavior is zealous attention to the few maintenance procedures for my 12-volt powersystem. The system has liberated me from so many of the problems of life in my wilderness home that the least I can do is make sure that it has every chance to operate efficiently. I spend a few minutes a week checking my batteries with a hydrometer, and generally checking out my power center and following up on any irregularities. In return I have enjoyed more than two years now of carefree country living with many of the city's comforts and conveniences for a fraction of the city's cost.

My neighbors and the manufacturers of some of the products we have installed have contributed to help me devise the routine maintenance list that follows. If you have followed the instructions in this book so far and if you did not take any shortcuts, the chances are good that you will have very little system trouble. But frankly, we have to thank some people, including myself, who did take shortcuts for some of these troubleshooting suggestions.

The total system requires three commonsense rules.

1. Always turn off your engines and disconnect the grounding wire from the battery post before working on any part of your system's wiring.

2. Conserve the energy you have so carefully provided by turning off lights and appliances, especially a power inverter, when they are not in use.

3. Respect the impact of any major change in the habitual amp-hour consumption and use of the system in your home. As you develop predictable charge-discharge cycles, your battery system virtually gains a memory of what you expect from it.

The starting-up phase, before your system has a chance to develop its memory, presents four special situations.

1. If a fuse blows the first time you reconnect the last battery wire, get out your volt/ohm meter and check the bad circuit (the one that blew the fuse) for reversed positive-negative polarity, a short circuit, or bad connections.

2. If no fuse blows but you simply have no electricity, check to be sure your car-to-house plug is wired correctly.

3. If you switch on something like your water pump and the lights dim considerably, you are experiencing voltage drop. You probably have a loose connection somewhere along the line, or perhaps you have too many lights and appliances on one circuit. You can always add more circuits, even another fuse box, when needed. A slight dimming of lights is normal.

4. After you have been at home for a while and the system suddenly and mysteriously fails, always check your outside connections first, especially if you have a house-to-car plug setup. A cat or a deer may have jumped across the wire and loosened the connection.

Prolong battery life with these six rules of preventive medicine.

1. Instead of one 12-volt battery, try two six-volt batteries in series. If a cell goes dead, you discard only half of your battery system.

2. Check the water level in the battery cells frequently,

and replenish the cells only with distilled water. Take frequent hydrometer readings.

3. Protect battery posts from corrosion by applying Vaseline or a commercial spray available at your service station or an auto supply dealer.

4. Never allow a piece of metal to touch positive (+) and negative (–) battery posts at the same time. The battery can short out and the cells can be destroyed.

5. Never smoke, light a match, or carry a fuel-ignited lamp near a battery, especially when charging. The flammable gases that build up during charging may be trapped inside an enclosed battery installation. Even though the enclosure has been vented to the outside, it is better to be safe than sorry.

6. Because of the gas build-up, be sure you enclose and vent your auxiliary batteries on-site. Explosions are rare, but are much less destructive when they are contained.

Keep the battery charged, according to these four tips, and it will continue to be one of the most dependable, trouble-free components of your system.

1. Start charging batteries when your volt/ohm meter reads 10 1/2 volts. Never let the batteries completely discharge.

2. If the picture on your 12-volt TV is growing progressively smaller and smaller, your batteries need recharging immediately.

3. About once a month, ask your service station to apply a booster charge of 60 amps or more over 30 to 45 minutes. The booster clears battery plates of lead sulphate and equalizes the charge in the battery cells.

4. If the car is your main power source and your current driving habits do not fully recharge your batteries, you may want to install a new alternator with a higher amperage output.

Your alternator is the key to keeping your battery fully charged, and the tension in your fan belt determines the amount of charge the alternator can supply in one drive. Check your fan belt frequently for looseness or wear, and follow the alternator troubleshooting checklist that follows. You might want to obtain a replacement alternator to have on hand just in case of failure. An extra fan belt is a must.

Condition	Correction

Alternator Light Stays On with Engine Running

(a) Belt is loose or broken.	Replace belt if necessary; tighten belt tension to specs.
(b) No alternator output.	Connect test lamp to alternator No. 1 terminal. If lamp lights dimly, alternator receives initial field current but has no output. Ground alternator field. If light does not brighten, remove and repair alternator. If light does brighten, the alternator is OK and voltage regulator is defective.

Alternator Light On with Ignition Off

(a) Positive diode shorted.	Replace diode.

Alternator Light Off with Ignition On but Engine Not Running

(a) Indicator bulb burned out.	Replace bulb and socket assembly.
(b) Blown fuse.	Replace fuse controlling indicator lights if all are off. If indicator lights do not come on with new fuse, check for an open indicator light feed wire.
(c) Open between bulb and ground in alternator.	Ground No. 1 terminal wire. If bulb lights, remove alternator for repair. If bulb still does not light, check further for an open circuit between bulb and correct.

Undercharged Battery

(a) Continuous small drain on battery.	Disconnect negative cable from battery. Connect test light between cable and battery post. If test lamp lights, trace and correct source of drain.
(b) Belt is loose or broken.	Replace belt if necessary; tighten belt to specs.
(c) Low alternator output.	Connect voltmeter across battery and record voltage reading. Set carburetor on high step of fast idle cam, start engine, and turn on all continuous use accessories. If voltage across battery reads lower than open circuit voltage just recorded, alternator current output is low. Check the voltage regulator first, if okay, remove and repair alternator.
(d) Low voltage regulator setting	Set up test as in (c) above but turn off all accessories. When engine is at operating temperature any reading between 13.5 and 15.0 volts indicates regulator is OK.
(e) High resistance in starting circuit or ignition resistor bypass.	Connect jumper from negative terminal of coil to ground to prevent starting engine. Connect voltmeter to positive coil terminal and ground. Crank engine. Any reading below 9.0 volts requires a further test. Check voltage across battery posts during cranking. If voltage reading is within 0.5-volt of that at coil, circuits are OK but battery is too low.

Battery Overcharged

(a) Shorted battery cell.	Perform battery load test to determine if battery cell is at fault. A shorted cell will result in a no-load battery voltage reading two volts low.
(b) Voltage regulator setting too high.	See Undercharged battery (c) above. Set up test as in (c) and perform as in (d). If voltage is over 15.0 volts, adjust or replace voltage regulator.

Based on *Petersen's Big Book of Auto Repair* by permission.

Fluorescent lighting can have a very long life, especially if you understand its operation. Notice how few instances require replacement of lamps in this checklist, courtesy of McLean Lighting.

Symptom	Possible Cause	Remedy
Lamps exhibit a swirling effect at turn on.	New lamps.	Goes away with time.
	Cold temperature.	Heat the home or the lamp.
Lamps take long time to come on, and/or low light output.	Low voltage from battery.	Check for weak or low battery (should be at least 9.6 volts) and battery should never get that low.
	Cold temperature.	Heat the home or the lamp.
	Lamp starting to blacken at end.	Turn lamp around.
Lamps blackened at both ends.	Prolonged use in low light condition.	Replace lamp.
Lamps blackened at both ends.	Lamp nearing end of its life.	Replace lamp.
Lamps won't come on.	No power.	Check other 12-volt devices.
	Blown fuse or circuit breaker.	Check fuses or circuit breakers.
	Lamps not properly installed in holder.	Check position of lamps in holder.
	Break in wiring.	Check for obvious broken or disconnected wires.
	Lamps burned out.	Replace lamps.

If a problem still exists after following the trouble shooting procedures, remove the fixture and, if possible, connect directly to a 12-volt DC source. If it still doesn't light follow the warranty procedures.

Avoid negligence with one final tip.

When you wake up one morning and nothing in your system works, check the car first. If it rolled away, your home will be unplugged. Resolve to set your emergency brake in the future.

As the years go by and the number of twelvoltaholics continues to grow, this repository of our folk remedies keeps increasing in length and scope. One reason for all this growth is the constantly expanding line of low-voltage appliances and accessories. When I encouraged you to start "making your choice" in Part One of this book, you really only started. Consumer demand will offer us many other choices in the near future.

Several manufacturers have introduced components that bring true concert-hall sound to 12-volt stereo systems. Laser Acoustics offers a 250-watt amplifier with three bands of equalization for a 12-volt system. Some newer 110-120-volt AC turntables, like the Technics model, actually convert AC current to DC to drive their motors. These turntables can be modified easily to play on 12-volts DC when you take the unit to a good radio or TV specialist. Also keep an eye out for new developments in catalogs from Panasonic, Craig, Isophon, and Hart Acoustics, among others.

In addition to the miniature digital calculators that have become so common, there are now little calculators that will print the figures on rolls of tape. They operate on tiny nickel-cadmium batteries that can be recharged by plugging them into your 12-volt system. All you need is a 6-, 7 1/2-, or 9-volt adapter, available for about $10 from an electronic supply house like Radio Shack or by mail order from me.

Adapters for different sizes of low-voltage equipment open up a whole range of new possibilities. Sophisticated video games and now home computers can also be adapted to plug into your powersystem.

Many elements in a home-computer system are inherently low-voltage designs. The keyboard console, video display, and casette motors can operate directly on 12-volt current with an adapter. With a power inverter, you can add a printer and disc-drive to your 12-volt system. A rotary inverter is best, but if your components are equipped with a switching regulator, a solid-state inverter can also help you build a complete electronic data center in the wilderness.

Have you seen all the small appliances in the stores which have that little black or gray box at the end of the cord? The little box is there to reduce the 110-120-volt AC electricity usually found in city offices and households. The appliances only require 12, 9, 7 1/2 or 6 volts to operate, and all we have to do is adapt them back to 12 volts.

There is now a catalog marketing company that specializes in low-voltage technology breakthroughs. Write JS & A Company, Northbrook, IL 60062.

You are probably becoming another low-voltage technology resource yourself. Now that you have designed and installed the powersystem, you cannot help but find new applications and solutions to problems that I did not encounter. If you have made other discoveries that make life in your wilderness home more comfortable, convenient, practical... please write to me.

One aspect of life here on my mountain constantly amazes me. The new breed of wilderness resident is incredibly resourceful. My neighbors are not just consumers in our society; they are great contributors to the society as well. Much of the material presented in this book resulted from conversations with, advice from, and working side by side with people who had faced some of the problems before me and helped immeasurably in the development of the solutions. Look around you, and I am sure you will find the willingness to share. Now you can join in.

Through my 12-volt enterprises and this publication, I hope to be able to serve as a clearinghouse for new ideas and products. Please notice that my interest does not stop with 12-volt technology. I am learning more and more about wind, water, and photovoltaic systems as well. There is an invitation at the end of the book for you to join my mailing list, if you have not already done so, to receive advance news of other new developments and publications.

Please keep in touch.

Glossary

AC, alternating current—electric current that reverses direction in a circuit at a given frequency, usually 50 to 60 cycles per second (50 to 60 Hz), or at an irregular ("wild") frequency.

Amp, ampere—unit of electric current produced by the force of one volt acting through the resistance of one ohm.

BTU, British Thermal Unit—conventional unit of heat in British and American measurement.

DC, direct current—electric current that flows in one direction only.

HP, horsepower—unit of power equal to 745.7 watts.

Hz, Hertz—cycles per second in the frequency of alternating current.

W, ohm—unit of electric resistance when one amp is produced by a potential of one volt across a conductor's terminals.

PVC, polyvinylchloride—chemical coating to protect plastics from weather or other deterioration.

RPM, revolutions per minute—units that measure the number of complete turns made by a rotary device on its axis.

Volt, unit of pressure, measured by the potential difference in a conductor when one amp flows against the resistance of one ohm, expending one watt.

Watt, unit of electric power equal to a current of one amp under the pressure of one volt.

National Electric Code

Parts One and Two frequently refer to the code as it relates to systems of 50 volts or less. The publisher of the code has not granted permission to reprint the related portions in their entirety. What follows is an outline for readers' guidance.

Copies of the code book, which covers all electrical standards, may be ordered from the publisher. Enclose $6.25 for postpaid shipment. Write to:

National Fire Protection Association
470 Atlantic Avenue
Boston, MA 02210

ARTICLE 720 governs systems of 50 volts or less. Related material may also be found in Articles 650, 725 and 760.

720-1 defines the scope of the article—50 volts or less, direct or alternating current.

720-2 refers to any hazardous installation and requires reference to Articles 500 through 517. A 12 volt home installation should not fall into this category.

720-3 is omitted in the code book.

720-4 states that wire or conductors smaller than # 12 copper cannot be used. Wire or conductors used to supply appliance branch circuits cannot be smaller than # 10 copper or equivalent.

720-5 holds that lampholders be rated at not less than 660 watts and must be used.

720-6 says receptacles cannot have a rating of less than 15 amps.

720-7 refers to kitchens, laundries or other areas where portable appliances will likely be used. Receptacles must be rated for not less than 20 amps.

720-8 is about overcurrent protection and must comply with Article 240.

720-9 tells about storage battery installations complying with Article 480.

720-10 is about grounding, which must comply with sections 250-5 (a) and 250-45. Does not usually apply to the 12 volt home application.

ARTICLE 240

   e. is about overcurrent protection.

      1. holds that circuit wiring must be protected from overcurrent by being properly rated as to wire size, being a copper conductor and fused to match the amperage capacity of the wire or conductor. For 12 volt systems the ampacity rating of copper wire or conductor is as follows:

      # 12 wire—20 amps maximum. Stranded or solid.

      # 10 wire—30 amps maximum. Stranded or solid.

      #8 wire—40 amps maximum. Stranded or solid.

2. refers to circuit breakers or fuses. They must be of an approved type, which includes automotive. Fuse holders must be clearly marked with maximum fuse size.

3. is about larger current-using appliances such as pumps, compressors, heater blowers and other, like motor-driven appliances. It says to follow the manufacturer's installation instructions. (Note: Motors controlled by automatic switching or manual latching must be installed according to section 430-32 (c), not included here.)

4. holds that overcurrent protective devices have to be installed in an accessible location on a vehicle not more than 18 inches from the point where the power supply connects to the vehicle circuitry. Outside the recreational vehicle, the device must be protected against physical damage and weather. The exception is an outside, low-voltage supply. It will be permitted if fused within 18 inches after entering the vehicle or after leaving a metal raceway.

f. says that switches must have a direct current rating not less than the connected load.

g. holds that lighting fixtures for low voltage must be approved.

h. states that cigarette lighter receptacles of 12 volts that will hold and energize a cigarette lighter must be installed in a non-combustible outlet box.

Article 480 is about the use of storage batteries.

480-1 states that this article applies to all stationary installations.

480-2 defines batteries as follows:

a. storage—a battery made up of one or more rechargeable cells of lead-acid, nickel-cadmium or other rechargeable electro-chemical types.

b. sealed cell or battery—one that has no way to add water, electrolyte, or to be able to measure the specific gravity. Individual cells can have a venting arrangement as described in 480-9 (b).

c. battery voltage is computed at 2.0 volts per cell for lead-acid types and 1.2 volts per cell for the alkali type.

480-3 states that any wiring and equipment which derives its power from a storage battery must meet the requirements of the National Electrical Code as it applies to wiring and equipment operating at the same voltage. The exception is stated in Article 800 regarding communications equipment.

480-4 is about grounding. The requirements of Article 250 apply.

480-5 is about the requirements for insulating batteries of 250 volts or less. This means cells connected together so that they produce and operate at a nominal battery voltage of not more than 250 volts.

a. says that vented lead-acid batteries—cells or batteries that have a sealed, non-conductive cover or heat-resistant materials—require no further insulation support.

b. says vented alkaline batteries—cells that have covers sealed—heat-resistant, non-conductive jars, will not need additional insulator support. Cells placed in jars of conductive material must be installed in trays of non-conductive material and no more than 20 cells (24 volts) may be in the series in any one tray.

c. says cells in rubber jars need no additional insulation if the nominal voltage of all cells connected in series does not produce more than 150 volts total. If the voltage exceeds 150, the batteries must be sectionalized into groups of 150 volts or less, and each group must have individual cells placed in trays or racks.

d. says that sealed cells or multi-compartment sealed batteries made of non-conductive, heat resistant material will require no more insulation. Batteries made with a conductive container around the cells must have insulation if a voltage is present between the battery container and the ground.

480-7 is about racks and containers needing to comply with (a) and (b) below.

a. states that racks are required to be of a rigid frame capable of supporting cells or trays, and must be:

1. metal to resist corrosive electrolyte and nonconductive insulation placed between the batteries and the metal, other than paint.

2. similar construction, i.e., fiber glass or other suitable non-metal materials.

b. holds that trays such as crates or shallow boxes, usually of wood or another non-conductive material, be made to be impervious to electrolyte or corrosion.

480-8 refers to locating batteries to conform as follows:

a. for ventilation—located to provide for proper diffusion and ventilation so that an explosive mixture cannot accumulate.

b. states that guarding of live parts must comply with Section 110-17. There is no requirement for 50 volts or less in that section.

480-9

a. tells that all vented cells must be equipped with a flame arrester made to prevent exploding of a cell if gas ignites within the cell by an external flame or spark under normal operating conditions. (Note: Most modern batteries are already designed this way by using plastic vent caps.)

b. states that batteries with sealed cells must be equipped with a pressure release vent or valve to prevent excessive pressure build-up, or be designed to prevent scattering the cell parts if a cell explosion occurs.

ARTICLE 551

551-3 is about low voltage systems.

  a. applies to the recreational vehicle manufacturer. The code says nothing specifically about home applications for 50 volts or less, but officials may impose the same requirements as for recreational vehicle manufacturers. In any case, many of these procedures and requirements are sound for home installations.

  b. is about wiring for low voltage.
    1. copper conductors (wire) must be used in all circuits.
    2. conductor or wire must be type HOT, SGT, or SGR, or type SXL, or the insulation rating of 60 degrees Centigrade or more, and a minimum wall thickness of 30 mils of thermoplastic insulation or the equivalent.
    3. for low voltage installations, single wire must be stranded type.
    4. says that low voltage insulated conductors must be surface-marked at no more than four-foot intervals. (Note: the rest of this section related to the marking requirements on the wire by the manufacturers of conductors.)

  c. refers to low voltage wiring methods.
    1. conductors must be protected against physical damage and secured. When insulated conductors are clamped to a wall or structure, the insulation must be supplemented by an additional wrapping of equivalent material, unless the conductor is in a jacketed material. Wiring must be kept away from sharp edges, moving parts and heat sources.
    2. conductors must be spliced or joined by approved splicing tools or by brazing, welding or soldering with a fusible metal or alloy. Soldered splices must be first joined so as to be mechanically and electrically secure without solder and then soldered. Splices, joints and free ends of conductors must be sealed with insulation equivalent to that on the conductor.
    3. battery and DC circuits must be completely separated by at least a 1/2-inch gap from circuits of a different power source (such as AC). Clamping, routing and equivalent means of insuring permanent separation are acceptable. Non-metallic sheathed cable may be used and considered to be adequate separation.
    4. ground terminals must be accessible for service. Ground terminal surfaces that make contact must be clean and free from oxide or paint, or be electrically connected with a cadmium, tin or zinc plated, external toothed lock washer or locking terminals. Ground terminal screws, rivets, bolts, nuts and lock washers must also be cadmium, tin or zinc plated with the exception of rivets that can be unanodized aluminum when attaching to an aluminum structure.

  d. is about installing batteries. They must be securely attached to the vehicle, be vaportight to the interior of a vehicle and ventilated with openings of a minimum of a 1.7 square-inch area at the top and the bottom. Batteries cannot be installed in a place where sparks or flame-producing equipment is located, except if they are installed in an engine generator compartment where the only charging source is the engine generator.

ARTICLE 551

551-4 deals with combining electrical systems.

  a. says that vehicles which have been code-wired for battery or direct current will be permitted to connect to a 115 volt source, providing that the entire wiring system and equipment fully conform to Part A requirements covering 115 volt electrical systems. Circuits powered from alternating current transformers shall not be connected to supply direct current appliances.

  b. talks about the 115 volt alternating side of a voltage converter (115 volts AC to low-voltage DC) must be wired in full conformity with Part A for 115-volt electrical systems. The exception is converters supplied as an integral part of an appliance. (Note: Part A has not been interpreted here because 115 volt AC wiring should only be done with consultation and help from a qualified electrician).

All converters and transformers must be listed for use in recreational vehicles and be designed or equipped with overheating protection. To determine the converter rating, apply this formula to the total connected load, including the average battery charging rate, of all 12 volt equipment: first 20 amperes of load at 100 percent, plus the second 20 amperes of load at 50 percent, plus all loads above 40 amperes at 25 percent.

  c. refers to all fixtures which have both 115-volt and low-voltage connections. They must be dual-voltage approved.

The information outlined here should help to insure a safe installation.

Metric Conversions

To convert	To	Multiply by
Acre	square meters	4047.0
Length in inches	millimeters	25.
Length in feet	centimeters	30.
Distance in miles	kilometers	1.6
Temperature in °Fahrenheit	°Celsius	0.56 (after subtracting 32)
Weight in pounds	grams	0.45
Volume in gallons	liters	3.8

Emergency Home Powersystem

As a significant by-product of research on my powersystem for wilderness homes, the Emergency Home Powersystem will provide limited power in a city home during public utility failures. With proper planning, this system will provide basic needs and comforts over several hours, or if operated more economically, it will serve over a number of days.

When power fails, you need not do anything. Your 12-volt lighting will remain on. You will have 12-volt outlets for radio and TV news about the emergency. You may operate the blower on a gas- or oil-burning furnace to stay warm during winter blackouts.

If you must operate your refrigerator, a microwave oven or some other 110-120-volt AC appliances, you can switch on a particular power inverter. The only exceptions are electric ranges, heaters, or air conditioners, which drain too much electricity.

You can create a simple system or a fully automatic variation by combining various components described in this book. By a few simple conversions to 12-volt plugs, you can add several of the small accessories described in Chapter 2. With adapters, you can plug in your 6-, 7 1/2- or 9-volt DC calculator, tape recorder, video games, computers, etc.

You will choose 12-volt batteries of the appropriate size to meet your anticipated demand. The placement of all the components is your choice, and the necessary wiring is fully explained earlier in this book.

However, you will become involved with your present 110-120-volt AC wiring to some extent, more so if you add a power inverter. Code requirements specify that a licensed electrician must asist you with these phases of your installation.

Your entire emergency system is a scale model of the powersystem described in this book. The following illustration and numbered instructions will show you how to assemble your model system.

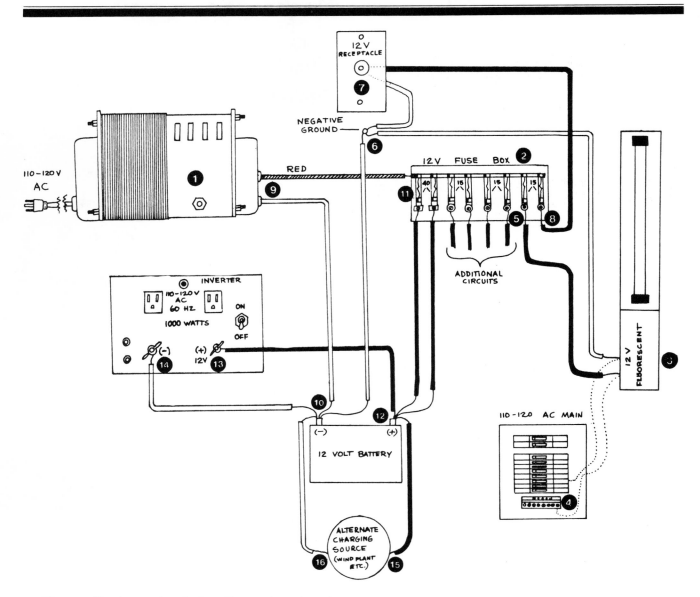

The portable charger described in Chapter 4 is a key element in this system. The charger (1) delivers 12-volt power to your fuse box (2), which has become an integral part of this charging system.

The fuses will keep circuits open for 12-volt fluorescent fixtures (3), which replace your existing 110-120-volt AC overhead lights. Remove the appropriate wires from the AC main (4), and reconnect them to the fuse box at points (5) and the negative ground bolt (6) as shown. (If your wiring is #12 or larger, no re-wiring is necessary.) You can also install 12-volt outlets (7) in each room, and connect their wiring to the fuse box at points (6) and (8). All of these 12-volt fixtures will be operable during normal utility power conditions.

Now for your emergency protection, the portable charger is connected from point (9) to the negative (−) terminal on the battery system (10) and via the fuse block at point (11) to the positive (+) terminal of the battery system (12).

For a fuller system, add the solid-state power inverter—1,000 watts continuous, 5,000 watts surge. Using #2 wire, make the positive (+) connection from the inverter (13) to the battery (12) and the negative (−) connection between points (14) and (10).

As you begin to enjoy the security and convenience of your alternative energy system, you may want to rely on it even more of the time. The addition of any of the charging alternatives we have discussed—wind, water, solar, or thermopile—is easy. Using the wire sizes appropriate to the generator you have chosen, simply make the positive (+) connections between points (15) and (12) and the negative (−) connection between points (16) and (10).

There is one lifesaving caution associated with use of the Emergency Home Powersystem. During a power failure, you must be certain that your power inverter cannot feed 110-120-volt AC energy back into the utility company's system. You could be responsible for serious injury to the people attempting to repair downed power lines. In other words, absolutely no connections can be made from the inverter AC output to any household outlet.

How much does it cost? The answer depends entirely on the degree of comfort and convenience you want, the number of lights and outlets you need, and the size of the battery system you require. The portable charging device is about $180, and you can take it from there. The appropriate inverter will cost about $1,000. Consult Chapter 2 for rough prices on the rest of your choices.

The total cost of a typical Emergency Home Powersystem will usually exceed the price of a traditional AC generator alone. But this system requires no fuel or oil changes and little maintenance. And later on, if the public utility companies are allowed to impose a surcharge on consumers during peak demand hours, you will save even more by simply switching to the battery power you have stored overnight. The more charging alternatives you add, the more power you will store. You do not have to wait for an emergency to gain the benefits of this home powersystem.

Index

Catalog Order Form

You may wish to send for my new 12-volt systems catalog, made possible in part by those special people, both consumers and builders, who have contributed to this unique information cooperative, people who are doing something about our energy problem.

If you do not wish to cut out this order form, please photocopy it or order on a separate sheet of paper. Please print your name and address.

Mail to:

Jim Cullen Enterprises, Inc.
The Wilderness Home Powersystem
Post Office Box 732
Laytonville, CA 95454

Dear Jim,

Enclosed is my check for $2.50 U.S.* to cover your costs. Please send me the current edition of your catalog, and add/do not add my name to your mailing list.

I am most interested in more information on the systems checked:

___New battery designs
___AC Generator/Battery charger
___Automotive Alternators
___Wind Plants
___Water Turbines
___Photovoltaic Arrays
___Thermopile Generators
___Emergency Home Powersystem
___New appliances and accessories

My home location is:

___City ___Beach

___Desert ___Mountains

Thank you.

Name_____

Address _____

_____zip_____

*Foreign air mail, please add $2.50 U.S. ($5.U.S. total).